LECKIE
the education publisher
for Scotland

T0173358

National 5
CHEMISTRY

Student Book

Tom Speirs and Bob Wilson

ISBN 9780008282080

Published by
Leckie
An imprint of HarperCollins*Publishers*
Westerhill Road, Bishopbriggs, Glasgow, G64 2QT
T: 0844 576 8126   F: 0844 576 8131
leckiescotland@harpercollins.co.uk   www.leckiescotland.co.uk

HarperCollins Publishers
Macken House, 39/40 Mayor Street Upper, Dublin 1,
D01 C9W8, Ireland

Special thanks to
Jennifer Richards (project management); Jouve (layout); Ink Tank (cover design); Jan Fisher (proofread); David Hawley (proofread); Alistair Coats (proofread), Jill Laidlaw (proofread)

Printed in Great Britain by Ashford Colour Press Ltd.

A CIP Catalogue record for this book is available from the British Library.

Acknowledgements
Whilst every effort has been made to trace the copyright holders, in cases where this has been unsuccessful, or if any have inadvertently been overlooked, the Publishers would gladly receive any information enabling them to rectify any error or omission at the first opportunity.

Leckie would like to thank the following copyright holders for permission to reproduce their material:

Cover: Shutterstock.com   Scottish Qualifications Authority for adapted questions from SQA Exam Papers; Area 1 opener: Lonely / Shutterstock.com; 1.1.1 TFoxFoto / Shutterstock.com; 1.1.2 Burben / Shutterstock.com; 1.1.3 CC STUDIO/SCIENCE PHOTO LIBRARY; 1.1.14 CHARLES D. WINTERS/SCIENCE PHOTO LIBRARY; 1.1.15 M&N / Alamy Stock Photo; 1.2.1 nexus 7 / Shutterstock.com; 1.2.2 Jeff J Mitchell/Getty Images; 1.2.3 Tena Rebernjak / Shutterstock.com; 1.2.7 Goodshot / Shutterstock.com; 1.2.25 Ryan McVay / Shutterstock.com; 1.2.26 Comstock / Shutterstock.com; 1.2.28 Ogwen / Shutterstock.com; 1.2.32 Alexander Gordeyev / Shutterstock.com; 1.2.34 photodisc / Shutterstock.com; 1.2.35 JAMES KING-HOLMES/ SCIENCE PHOTO LIBRARY; 1.2.36 PETER MENZEL/SCIENCE PHOTO LIBRARY; 1.2.37 NASA/SCIENCE PHOTO LIBRARY; 1.2.41 Tobik / Shutterstock.com; 1.2.41 Nneirda / Cherniga Maksym; 1.4.1 g215 / kwhi02 / Pixel Embargo / kwhi02 / Shutterstock.com; 1.4.2 MARTYN F. CHILLMAID / SCIENCE PHOTO LIBRARY; 1.4.3 SCIENCE PHOTO LIBRARY; 1.4.4 SHEILA TERRY/SCIENCE PHOTO LIBRARY; 1.4.6 Blazius / Shutterstock.com; Area 2 opener: Lonely / Shutterstock.com; 2.5.3 NASA Earth Observatory image by Jesse Allen and Robert Simmon/VIIRS/Suomi NPP; 2.5.7 Natursports / Shutterstock.com; 2.5.9 Tyler Olson / Shutterstock.com; 2.6.4 CGissemann / Shutterstock.com; 2.6.6 JIM VARNEY/SCIENCE PHOTO LIBRARY; 2.6.12 HANS REINHARD/ OKAPIA/SCIENCE PHOTO LIBRARY; 2.6.13 T.L. / Shutterstock. com; Area 3 opener: Lonely / Shutterstock.com; 3.7.1 Lee Prince / Shutterstock.com; 3.7.2 SHEILA TERRY/SCIENCE PHOTO LIBRARY; 3.7.3 London 2012; 3.7.4 www.moneysideoflife.com; 3.7.5 Daniel CD / Shutterstock.com; 3.7.6 farbled / Shutterstock. com; 3.7.7 benjasanz / Shutterstock.com; 3.7.10 Kekyalyaynen / Shutterstock.com; 3.7.12 Digital Vision / Shutterstock.com; 3.7.15 ANDREWLAMBERT PHOTOGRAPHY/SCIENCE PHOTO LIBRARY; 3.7.16 CHARLES D. WINTERS/SCIENCE PHOTO LIBRARY; 3.7.17 SHEILA TERRY/SCIENCE PHOTO LIBRARY; 3.7.22 Pure Energy Centre; 3.7.23 EQUINOX GRAPHICS/SCIENCE PHOTO LIBRARY; 3.7.24 Transport for London; 3.7.27 ArielMartin / Shutterstock. com; 3.8.1 London 2012;3.8.2 London 2012; 3.8.3 Nike, Inc.; 3.8.4 Felt Bicycles; 3.8.9 The Advertising Archives; 3.8.10 yotrak / Shutterstock.com; 3.8.25 Rodolfo Arpia / Shutterstock.com; 3.8.13 Elke Wetzig / Shutterstock.com; 3.8.14 Vegware / Shutterstock. com; 3.8.17 PASCAL GOETGHELUCK/SCIENCE PHOTO LIBRARY; 3.8.18 PASCAL GOETGHELUCK/SCIENCE PHOTO LIBRARY; 3.9.5 EMILIO SEGRE VISUAL ARCHIVES/AMERICAN INSTITUTE OF PHYSICS/SCIENCE PHOTO LIBRARY; 3.9.9 MARTYN F. CHILLMAID/SCIENCE PHOTO LIBRARY; 3.9.13 nthg / Shutterstock.com; 3.9.13 Elsa Naumann/Sahara Forest Project; 3.10.3 Photos.com; 3.10.4 Marie Curie Cancer Care; 3.10.5 Whitepaw (talk); 3.10.6 Getty Images / Thinkstock Images; 3.10.9 alterfalter / Shutterstock.com; 3.10.11 P.PLAILLY/E.DAYNES/ SCIENCE PHOTO LIBRARY; 3.10.13 SCOTT CAMAZINE/SCIENCE PHOTO LIBRARY; 3.11.2 Chronicle / Alamy Stock Photo ; 3.11.5 EUROPEAN SPACE AGENCY/DENMAN PRODUCTIONS/SCIENCE PHOTO LIBRARY; 3.11.6 KNMI/IASB/EUROPEAN SPACE AGENCY/SCIENCE PHOTO LIBRARY; 3.11.7 VICTOR DE SCHWANBERG / SCIENCE PHOTO LIBRARY; 3.11.9 Jon Daughtry (dedass.com); 3.11.10 Rob van Esch / Shutterstock.com; 3.11.11 Hiroshi Ichikawa / Shutterstock.com; 3.11.5 Anucha Cheechang / Shutterstock.com; 3.11.16 ggw / Shutterstock.com; 3.11.13 ANDREW LAMBERT PHOTOGRAPHY/SCIENCE PHOTO LIBRARY; 3.11.20 Science Photo Library; 3.11.22 ANDREW LAMBERT PHOTOGRAPHY / SCIENCE PHOTO LIBRARY; 3.11.24 NASA, 3.11.25 NASA; 3.11.26 Digital Vision; 3.11.27 ANDREW LAMBERT PHOTOGRAPHY/SCIENCE PHOTO LIBRARY

Teacher notes
- *https://collins.co.uk/pages/scottish-curriculum-free-resources*

Introduction

About this book

This book provides a resource to practise and assess your understanding of the chemistry covered for the National 5 qualification. The book has been organised to map to the course specifications and is packed with examples, explanations, activities and features to deepen your understanding of chemistry and help you prepare for the final exam.

Features

YOU SHOULD ALREADY KNOW:

Each chapter begins with a list of topics you should already know before you start the chapter. Some of these topics will have been covered at curriculum levels 3 and 4 or at National 4, while others will depend on preceding chapters in this book.

> **You should already know**
>
> - During a chemical reaction, one or more of the following may be seen: a colour change; bubbles of gas; a solid formed; an energy change.

LEARNING INTENTIONS:

After the list of things you should be familiar with, there is a list of the topics covered in the chapter. This tells you what you should be able to do when you have worked your way through the chapter.

> **Learning intentions**
>
> In this chapter you will learn about:
>
> - Chemistry in action: the importance of rate of reaction.

EXAMPLE

New topics involving calculations are introduced with at least one worked Example, which shows how to go about tackling the questions and activities. Each Example breaks the question and solution down into steps, so you can see what calculations are involved, what kind of rearrangements are needed and how to work out the best way of answering the question.

Example 1.1.1

Calculate the average rate of reaction over the first 20 seconds.

Worked answer:

Average rate of reaction

$$= \frac{\text{volume at 20s} - \text{volume at 0s}}{\text{time interval}}$$

$$= \frac{49 - 0}{20 - 0}$$

$$= 2\cdot45 \ cm^3 s^{-1}$$

ACTIVITIES

Each chapter features a number of activities, to work on individually, in pairs or small groups. Activities have been carefully designed to test your understanding of the topic and provide experiences to deepen your understanding of the concepts and techniques involved.

> ### GO! Activity 1.1.1: Paired activity
>
> You can discuss these activities with a partner or complete them on your own. Make sure you check your answers.
>
> 1. Potassium nitrate is used in the propellant rather than sodium nitrate because sodium nitrate absorbs water more readily than potassium nitrate.
>
>   Suggest why there is a concern about the reactants absorbing water.
> 2. The particle size of the chemicals in the propellant must be between 5 and 20 microns (thousandth of a millimetre).

STEP BACK IN TIME

Most chapters have Step Back in Time features that add context to the chemistry topics. They help to show the history of the chemistry and the people involved in the major discoveries and developments in the science.

STEP BACK IN TIME: THE DISCOVERY OF STABLE ISOTOPES

The existence of isotopes had been proposed to explain aspects of radioactivity by Frederick Soddy when working at Glasgow University at the beginning of the twentieth century.

SPOTLIGHT

Most chapters highlight contemporary applications of chemistry – illustrating the real-world uses of chemistry and how it relates to industrial, commercial and everyday life.

SPOTLIGHT ON FORENSIC SCIENCE

Identifying the quantity of specific isotopes in a material can be used to work out where it has come from. A recent development in forensic science is the isotopic analysis of hair strands. Hair grows at an average rate of 1cm a month.

HINTS, WORD BANKS AND REMEMBER!

Where appropriate, Hints are given in the text to help give extra support. Word banks help to secure the terms and phrases used in the chemistry course to help you to become familiar with them and their usage.

> ### 📖 Word bank
>
> • **Average rate**
> The change in the quantity of reactant or product over time.

> ### 🔍 Hint
>
> As a reaction progresses the reactants get used up so the number of particle collisions decreases. Eventually at least one of the reactants gets used up and the reaction stops.

> ### ⚠ Remember!
>
> Ion-electron equations can be found in the SQA data booklet.

MAKE THE LINK

Chemistry is not a subject that exists in isolation! Where appropriate, link to other subject areas show the links that exist between different disciplines.

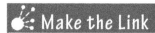

Make the Link

You will already know from maths how to read graphs to obtain information.

LEARNING CHECKLIST

Each chapter closes with a summary of learning statements showing what you should be able to do when you complete the chapter. You can use the Learning checklist to check you have a good understanding of the topics covered in the chapter. Traffic light icons can be used for self-assessment.

Learning checklist

In this chapter you have learned:

- The rate of a chemical reaction can be monitored by measuring the volume of gas produced over time.

- The rate of a chemical reaction can be monitored by measuring the loss in mass of reactants over time.

ASSESSMENTS

Assessments are provided for each of the areas. These assessments contain a number of exam-style questions and are good preparation for the final exam.

TEACHER NOTES

Teacher notes, giving guidance and suggestions for teaching the topics covered in each chapter are provided online at https://collins.co.uk/pages/scottish-curriculum-free-resources

1. **Rates of reaction**
   - Chemistry in action
   - Monitoring the rate of reaction
   - Calculating average rates of reaction

2. **Atomic structure and bonding**
   - Atomic structure
   - Periodic table
   - Nuclide notation
   - Isotopes
   - Relative atomic mass (RAM)
   - Covalent bonding and the shapes of molecules
   - Structure and properties of covalent substances
   - Structure and properties of ionic compounds

3. **Formulae and reaction quantities**
   - Formulae of elements and compounds
   - Chemical equations
   - Chemical formulae for compounds containing ions with more than one atom
   - Balancing chemical equations
   - Gram formula mass and the mole
   - Calculations involving mass into moles and moles into mass
   - Connecting mass, volume of solutions, concentration and moles
   - Calculating quantities from balanced equations
   - Calculating percentage composition

4. **Acids and bases**
   - Acids and bases all around
   - The dissociation of water molecules
   - pH and hydrogen ion concentration
   - Diluting acids and alkalis
   - What reacts in a neutralisation reaction?
   - Volumetric titrations
   - Preparing soluble salts

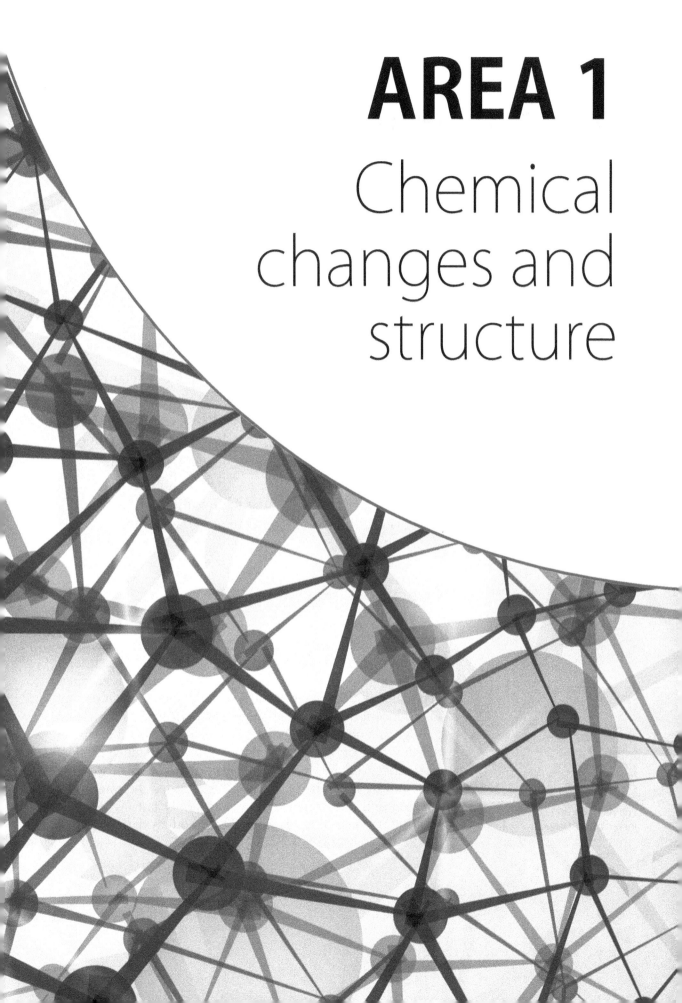

AREA 1
Chemical changes and structure

1 Rates of reaction

You should already know

- During a chemical reaction, one or more of the following may be seen: a colour change; bubbles of gas; a solid formed; an energy change.

Learning intentions

In this chapter you will learn about:

- Chemistry in action: the importance of rate of reaction.
- Increasing the rate of a reaction by increasing the concentration, decreasing the particle size, raising the temperature of the reactants and adding a catalyst.
- Monitoring change in rate of a chemical reaction by measuring the volume of gas produced and the change in mass of reactants over time.
- How to draw and interpret graphs of volume of gas produced and change in mass of reactants against time.
- Calculating the average rate of reaction using the information from graphs of volume of gas produced, loss in mass and change in concentration against time.

Chemistry in action

Industry produces a vast range of substances, from bulk chemicals such as sodium hydroxide (caustic soda) and ammonia, to medicines, chemicals used in the electronics industry and substances of biological origin such as vaccines and material for biofuels. Whatever the product, it is important to produce it in the most economic way. This can be achieved through an understanding of the factors that control a chemical reaction. Not only do we need to consider how much product a reaction gives and what the energy costs are, we need to know how **quickly** it can be produced. This is called the **rate** of the reaction and rates can vary from incredibly fast (as in a gas explosion) to very slow (as when a car's bodywork rusts).

Figure 1.1.1: *An explosion is an incredibly fast chemical reaction*

Figure 1.1.2: *Rusting is a very slow chemical reaction*

Airbag protection systems – saving lives with explosions

Airbag protection systems are commonplace in most road vehicles and require the bags to be inflated within 60–80 milliseconds (thousands of a second) of the car's sensors detecting the likelihood of an impact. This means that the gas which fills the bag must be produced in an explosively fast chemical reaction. In many airbag systems two reactions take place. The first, known as the initiator, produces enough energy to ignite a solid propellant – the chemical which produces the gas for the airbag. The gas produced is nitrogen, which is extremely unreactive and non-toxic.

Many of the airbag systems contain a mixture of sodium azide (NaN_3), potassium nitrate (KNO_3), and silicon dioxide (SiO_2) in the propellant. Within about 40 milliseconds of impact, all these components react in three separate reactions that produce nitrogen gas. The reactions, in order, are as follows:

1. The decomposition of sodium azide, which produces sodium metal and nitrogen gas:

    $$2NaN_3 \rightarrow 2Na + 3N_2 \text{ (g)}$$

2. The sodium metal is removed by reacting it with the potassium nitrate and silicon dioxide. More nitrogen gas is also produced:

    $$10Na + 2KNO_3 \rightarrow K_2O + 5Na_2O + N_2 \text{ (g)}$$

3. In the final reaction silicon dioxide is used to eliminate the potassium oxide and sodium oxide produced in reaction (2), because Group 1 metal oxides are highly reactive. These products react with silicon dioxide to produce a silicate glass which is a harmless and stable compound:

$$K_2O + 5Na_2O + 6SiO_2 \rightarrow K_2O_3Si + 5Na_2O_3Si \text{ (silicate glass)}$$

Chemists are carrying out research to find alternative propellant compounds. It is hoped that non-azide reagents will produce less reactive by-products and will also have a lower combustion temperature.

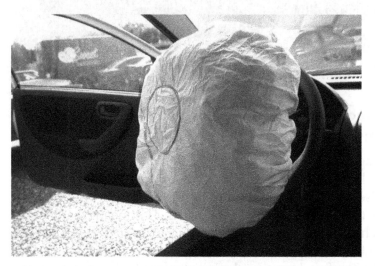

Figure 1.1.3: *Inflating an airbag relies on the rapid production of gas from an explosively fast chemical reaction*

Activity 1.1.1: Paired activity

You can discuss these activities with a partner or complete them on your own. Make sure you check your answers.

1. Potassium nitrate is used in the propellant rather than sodium nitrate because sodium nitrate absorbs water more readily than potassium nitrate.

   Suggest why there is a concern about the reactants absorbing water.

2. The particle size of the chemicals in the propellant must be between 5 and 20 microns (thousandth of a millimetre).

   Comment on the need for the particle sizes to be so small.

3. Suggest why it is essential to remove the sodium produced in reaction 1.

Monitoring the rate of reaction

The rate of chemical reactions can be monitored in a number of ways. A simple way is to measure the **volume of gas** produced in a reaction over time, then draw a graph of the results. A suitable reaction is marble chips (pieces of calcium carbonate) reacting with hydrochloric acid. Carbon dioxide gas is produced.

Both of the arrangements shown in Figure 1.1.4 and Figure 1.1.5 can be used – bubbling through water is the most common method used in the laboratory but using a syringe gives more accurate measurements and can be used for any gas.

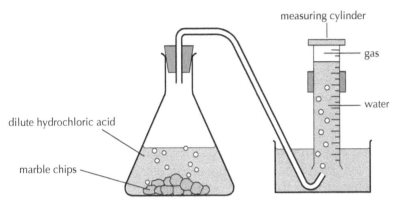

Figure 1.1.4: *The volume of gas produced during a chemical reaction can be measured by collecting the gas over water*

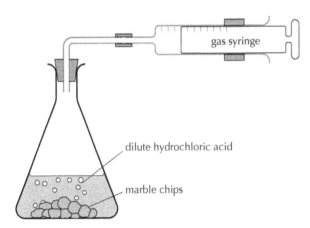

Figure 1.1.5: *The volume of gas produced during a chemical reaction can be measured by collecting the gas in a syringe*

The **loss in mass** over time can also be used to monitor the progress of a reaction. The arrangement shown in Figure 1.1.6 could be used. A computer connected to an electronic balance is used to collect data from the experiment.

Table 1.1.1: *Volume of gas collected over time*

Time (min)	Gas volume (cm³)
0	0
1	20
2	35
3	45
4	50
5	52
6	53
7	53
8	53

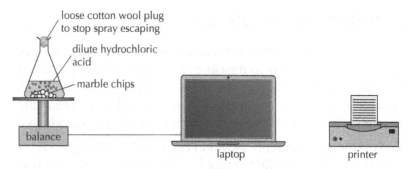

Figure 1.1.6: *Set-up for experiment measuring loss in mass*

As the carbon dioxide is produced, it escapes into the air and so the mass of the reaction mixture decreases.

Both experiments do the same thing – they measure the amount of carbon dioxide produced over time.

The rate of reaction was monitored in a number of experiments using marble and hydrochloric acid. The variables concentration, temperature and particle size were changed but the volume of acid used and the mass of marble used were the same in each experiment and all of the marble reacted. This means the total volume of gas produced in each experiment is the same. For each variable there were two experiments, so that the effect of changing the variable could be measured.

Changing concentration

Experiment 1

The **volume of gas** collected every minute was measured using the arrangement shown in Figure 1.1.4 Table 1.1.1 shows the results. A graph of volume of gas against time was drawn.

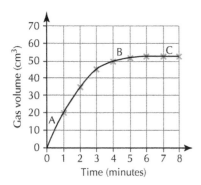

Figure 1.1.7: *Volume of gas collected over time*

Look at sections A, B and C of Figure 1.1.7.

- At A, near the start of the reaction: the graph is almost a straight line with a fairly steep slope. The gas was being produced quickly. The reaction was **fast**.

- At B, towards the end of the reaction: the graph starts to level off. The gas was not being produced as quickly. The reaction was **slowing down**.

- At C, at the end of the reaction: the graph has completely levelled off. All of the marble has reacted. No more gas was being produced. The reaction has **stopped**.

Experiment 2

More concentrated hydrochloric acid was reacted with the marble chips. The graph shown in Figure 1.1.8 was obtained – the original graph shown in Figure 1.1.7 is included to make it easier to compare the results.

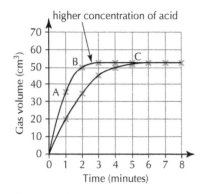

Figure 1.1.8: *Volume of gas collected over time, with different acid concentration*

- At A, the graph is again a straight line but with a much steeper slope than with the lower concentration of acid. The reaction was much **faster**.

- At B, the graph starts to level off more quickly than in the first experiment. The gas was not being produced as quickly.

- At C, the graph has completely levelled off at the same volume as the first experiment. This is because, although the concentration of the acid has changed, the mass of marble reacting is the same. The gas may have been produced more quickly but the same final volume of gas was produced in both experiments.

Changing temperature

Experiment 3

Experiment 1 was repeated but instead of collecting the gas, the **loss in mass** was measured over time using the arrangement shown in Figure 1.1.6. Table 1.1.2 shows the results.

The graph is similar to the one obtained in experiment 1 and the shape of the graph at points A, B and C can be explained in the same way (Figure 1.1.9).

Experiment 4

When the experiment was carried out with the acid at a **higher temperature** the graph shown in Figure 1.1.10 was obtained. The graph from Figure 1.1.9 is included to make a comparison easier.

- At A, the graph is again a straight line but with a much steeper slope than with the acid at room temperature. The reaction was much **faster**.

- At B, the graph starts to level off more quickly than in the first experiment. Mass was not being lost as quickly.

- At C, the graph has completely levelled off at the same mass as the experiment with the acid at room temperature. This is because, although the temperature of the acid has changed, the mass of marble reacting was the same. Mass was lost more quickly at the higher temperature but the same amount of mass was lost overall in both experiments.

Changing the particle size

Experiment 1 was repeated using **lumps** of marble. Experiment 2 used **powdered** marble. All other variables were kept the same. The results were monitored and the graphs are shown in Figure 1.1.11.

Table 1.1.2: *Loss in mass of marble over time*

Time (min)	Loss in mass (g)
0	0.000
1	0.037
2	0.064
3	0.082
4	0.092
5	0.095
6	0.097
7	0.097
8	0.097

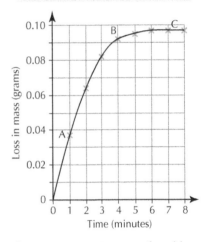

Figure 1.1.9: *Loss in mass of marble over time*

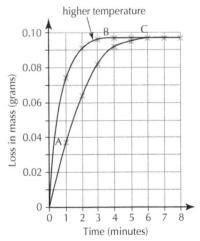

Figure 1.1.10: *Loss in mass of marble over time, at different temperatures*

The shape of each graph can be explained as in the previous experiments. The slope of the graph with the powdered marble is much steeper indicating that the reaction was faster when the marble is powdered as it has a larger surface area for the acid to react with.

End-point and quantities produced and reacting

The end-point of a reaction is the time at which the graph levels out (goes flat). This means no more gas is being produced – the reaction has stopped.

Look at Figure 1.1.11, which shows the volume of gas produced when powdered marble and lumps of marble react with hydrochloric acid.

In each reaction, the volume of gas produced at the endpoint is $80\,cm^3$. The volumes are the same because the same mass of marble is reacting in each experiment.

However, the time taken to reach the end-point in each reaction is different:

- marble powder: approximately 1.6 min

- marble lumps: approximately 3.5 min.

The marble powder reaches the end-point more quickly (it takes less time), so it is the faster reaction.

The volume of gas produced at other times can also be obtained from the graph.

- After 2 min, $60\,cm^3$ of gas is produced with marble lumps.

- The time taken to produce $30\,cm^3$ of gas using marble powder is 0.5 min.

Look at Figure 1.1.9, which shows loss in mass when marble reacts with hydrochloric acid. The end-point of the reaction occurs after 6 min and the loss in mass of reactants is just below 0.10 g (0.097 g).

Sometimes the change in the total mass of the reaction flask and its contents over time is measured and a graph plotted. Figure 1.1.12 shows the type of curve obtained.

Although the shape of the graph is different to the other graphs in this section, it can be interpreted in the same way.

- At A, the graph is a steep straight line, showing that the reaction is fast. The reactants are being used up quickly.

- At B, the graph is less steep, showing that the reaction is slowing down because the reactants are being used up.

- At C, the graph has levelled off. At least one of the reactants is completely used up and the reaction has stopped. This is the end-point of the reaction.

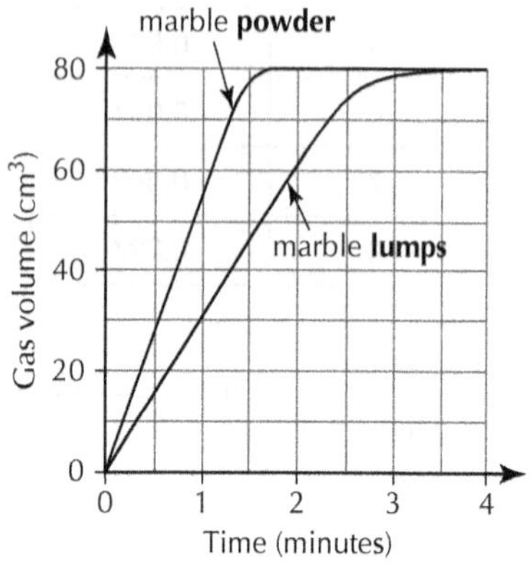

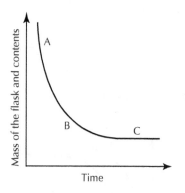

Figure 1.1.11: *Volume of gas collected over time*

Figure 1.1.12: *Change in total mass over time*

Adding a catalyst

Many chemical reactions can be speeded up by adding a **catalyst**. For example, hydrogen peroxide slowly produces oxygen gas when left in the bottle. However, when a catalyst is added, the oxygen is produced much more quickly.

The concentration, volume and temperature of the hydrogen peroxide are kept the same.

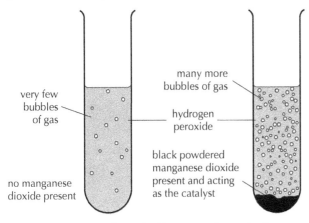

very few bubbles of gas

many more bubbles of gas

hydrogen peroxide

black powdered manganese dioxide present and acting as the catalyst

no manganese dioxide present

Figure 1.1.13: *The effect of adding a catalyst to hydrogen peroxide*

Figure 1.1.14: *Dramatic effect of adding a catalyst to hydrogen peroxide in the laboratory*

Conclusion: Adding a catalyst increases the rate of a reaction. The catalyst does not get used up and is chemically unchanged at the end of the reaction.

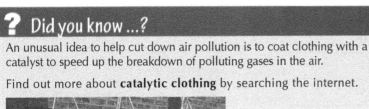

❓ Did you know ...?

An unusual idea to help cut down air pollution is to coat clothing with a catalyst to speed up the breakdown of polluting gases in the air.

Find out more about **catalytic clothing** by searching the internet.

Figure 1.1.15: *The dye used in denim jeans is good at absorbing the catalyst which can break down atmospheric pollutants*

🔬 Make the Link

• **Catalysts in industry**
Catalysts are widely used in industry. They speed up chemical processes and reduce energy costs.

GO! Activity 1.1.2

1. Powdered zinc metal can be reacted completely with hydrochloric acid. One of the products is hydrogen gas. The volume of hydrogen produced can be measured and the results used to see how the rate of reaction changes over time.

   (a) Describe how you could carry out an experiment to measure the volume of gas produced over time. Include a labelled diagram.

   (b) The rate-of-reaction graph produced from the results of an experiment like the one you described in part (a) is shown.

      (i) Is the reaction faster at point A or point B?

      (ii) What is it about the line on the graph that helped you to answer part (b) (i)?

      (iii) What has happened to the reaction at point C on the graph?

   (c) What volume of hydrogen gas was produced after 3 minutes?

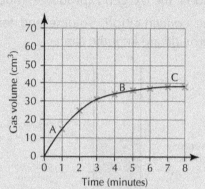

Volume of gas produced when powdered zinc metal reacts with hydrochloric acid

   (d) (i) Sketch the graph above and add another line to show what happens when the reaction is carried out at a higher temperature.
      (You do not need to use graph paper or include the scales on the graph.)

      (ii) Explain the shape of this graph compared to the original.
      (Hint: Experiment 4 on page 15 will help you answer this question.)

   (e) The table below shows results for a similar reaction.

Time (min)	Volume of hydrogen (cm³)
0	0
1	15
2	25
3	31
4	34
5	34
6	

   Predict the volume of gas collected after 6 minutes.

2. The rate of a chemical reaction can be monitored by measuring the loss in mass over time.

   (a) Describe how you could carry out an experiment to measure the loss in mass over time, when marble chips react with hydrochloric acid. Include a labelled diagram.

(b) The graphs below show loss in mass over time when the same mass of powdered chalk and lumps of chalk react with hydrochloric acid. All the chalk reacts in each reaction.

(i) Which graph, X or Y, shows the reaction with powdered chalk?

(ii) Explain your answer to part (b) (i).

(c) How many minutes does it take for the reaction represented by graph X to finish?

(d) From the graphs, work out the total loss in mass in both experiments.

(e) Suggest why the total loss in mass is the same in each experiment.

(f) If a more concentrated acid had been used than the one used to produce graph X, how would the shape of the graph near the start of the reaction compare to graph X?

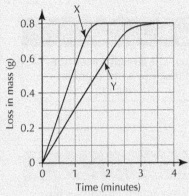

Loss in mass when powdered chalk and lumps of chalk react with hydrochloric acid

Calculating average rates of reaction

The average rate of a chemical reaction is the change in the quantity of reactant or product over time and can be calculated using data collected from experiments carried out in the laboratory.

A suitable reaction is marble chips (containing calcium carbonate) reacting with hydrochloric acid:

calcium carbonate + hydrochloric acid → calcium chloride + water + carbon dioxide

$$CaCO_3(s) \quad + \quad 2HCl(aq) \quad \rightarrow \quad CaCl_2(aq) \quad + H_2O(\ell) \quad + \quad CO_2(g)$$

(1) Measuring the change in volume of gas produced during the reaction

The rate at which carbon dioxide gas is given off can be measured by collecting the gas and measuring the volume at fixed time intervals. Collecting the gas by displacement of water is the most commonly used method in the laboratory but using a syringe gives more accurate measurements. (See Figures 1.1.4 and 1.1.5 on page 13.)

Figure 1.1.16 shows how the volume of gas changes over the course of the reaction – the gas volume readings were taken every 10 seconds.

> 📖 **Word bank**
>
> • **Average rate**
> The change in the quantity of reactant or product over time.

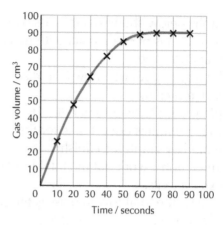

Figure 1.1.16: *Graph of volume of gas produced over time*

Make the Link

You will already know from maths how to read graphs to obtain information.

Hint

As a reaction progresses the reactants get used up so the number of particle collisions decreases. Eventually at least one of the reactants gets used up and the reaction stops.

The average rate of reaction is calculated by dividing the change in volume over a measured time interval:

Average rate = change in volume / change in time

The information required is obtained from the graph.

Example 1.1.1

Calculate the average rate of reaction over the first 20 seconds.

Worked answer:

Average rate of reaction

$$= \frac{\text{volume at 20s – volume at 0s}}{\text{time interval}}$$

$$= \frac{49 - 0}{20 - 0}$$

$$= 2 \cdot 45 \ \text{cm}^3 \, \text{s}^{-1}$$

Since rate here is a measure of the change in volume of the gas over time the unit of rate is **$\text{cm}^3 \text{s}^{-1}$** (cubic centimetres per second).

(2) Measuring the loss in mass during a reaction

The rate at which carbon dioxide is given off is obtained by measuring the loss in mass of the reactants at regular time intervals. As the gas is produced, it is released into the air so the mass of the reaction mixture and flask decreases. Figure 1.1.6 on page 14 shows a suitable arrangement for carrying out the experiment.

The graph shows how the mass changes over the course of the reaction – the mass readings were taken every 10 seconds.

The average rate of reaction is calculated by dividing the mass loss by the time interval:

Average rate = change in mass/change in time

The information required is obtained from the graph.

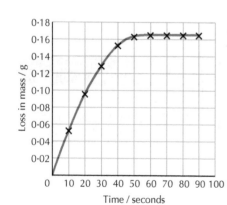

Figure 1.1.17: *Graph of loss of mass over time*

Calculate the average rate of reaction between 20 and 30 seconds.

Example 1.1.2

Worked answer:

Average rate of reaction

$$= \frac{\text{mass loss at 30s} - \text{mass loss at 20s}}{\text{time interval}}$$

$$= \frac{0{\cdot}128 - 0{\cdot}098}{30 - 20}$$

$$= \frac{0{\cdot}03}{10}$$

$$= 0{\cdot}003 \text{ g s}^{-1}$$

Since rate here is a measure of the mass loss over time the unit of rate is **g s⁻¹** (grams per second).

(3) Measuring the change in the concentration of the acid during a reaction

The graph shows the change in concentration of the acid over time:

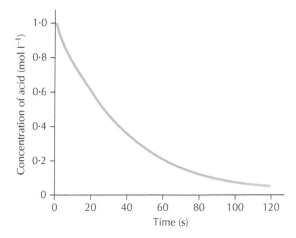

Figure 1.1.18: *Graph of concentration of acid against time*

The average rate of reaction is calculated by dividing the change in concentration by the time interval:

Average rate = change in concentration/change in time

The information required is obtained from the graph.

Example 1.1.3

Calculate the average rate of reaction between 20 and 40 seconds.

Worked answer:

Average rate of reaction

$$= \frac{\text{concentration at 20s} - \text{concentration at 40s}}{\text{time interval}}$$

$$= \frac{0.64 - 0.38}{40 - 20}$$

$$= \frac{0.26}{20}$$

$$= 0.013 \text{ mol } l^{-1}s^{-1}$$

(Note that because the concentration decreases with time the concentration at the higher time is subtracted from the concentration at the lower time.)

Since rate here is a measure of the change in concentration over time the unit of rate is **mol l^{-1} s^{-1}** (moles per litre per second).

GO! Activity 1.1.3: Paired activity

You can discuss the activities with a partner or complete them on your own. Make sure you check your answers.

1. Look at the graph of volume of gas against time (Figure 1.1.16).
   (a) Calculate the average rate of reaction between 20 and 40 seconds.
   (b) Compare your answer in (a) with the average rate over the first 20 seconds (Example 1.1.1) and explain the difference in average rates over the two time intervals.
   (c) State the time at which the reaction stopped and explain how you got this value.

2. Look at the graph of loss of mass against time (Figure 1.1.17).
   Calculate the average rate of reaction over the first 10 seconds.

3. Look at the graph of concentration against time (Figure 1.1.18).
   Calculate the average rate of reaction over the first 20 seconds.

4. The table shows the loss in mass in a reaction of magnesium with hydrochloric acid.

Time/min	Loss in mass/g
0	0.000
1	0.037
2	0.064
3	0.082
4	0.092
5	0.095
6	0.097
7	0.097
8	0.097

   (a) Use the information in the results table to sketch a graph of loss in mass against time (you do not have to use graph paper).
   (b) The graph levels out at around six minutes.
   Explain why this is.

Learning checklist

In this chapter you have learned:

- The rate of a chemical reaction can be monitored by measuring the volume of gas produced over time.

- The rate of a chemical reaction can be monitored by measuring the loss in mass of reactants over time.

- The slope of a rate-of-reaction graph is steep near the start of a reaction, which indicates that the reaction is fast at this point.

- The slope of a rate-of-reaction graph becomes less steep when a reaction is slowing down.

- The slope of a rate-of-reaction graph eventually levels off because the reaction has stopped.

- The average rate of reaction

    = change in volume of gas/change in time

    or

    = loss in mass/change in time

    or

    = change in concentration/change in time

- The data needed to calculate average rate can be obtained from graphs of volume of gas produced, loss in mass and change in concentration, all measured over time.

2 Atomic structure and bonding

Learning intentions

In this chapter you will learn about:
- The particles which make up an atom and how they are arranged.
- Atomic number and mass number.
- Nuclide notation for an atom and ion.
- Isotopes.
- Working out relative atomic mass (RAM).
- Covalent bonding and the shapes of molecules.
- Structure and properties of covalent substances.
- Structure and properties of ionic compounds.

◄€ SPOTLIGHT ON SCIENTIFIC RESEARCH: SEARCHING FOR THE 'GOD PARTICLE'

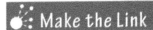

Make the Link

Theories put forward to explain the origins of the universe are discussed in both science and religious studies.

The atom was thought to be the smallest possible particle until protons, electrons and neutrons were discovered in the first part of the twentieth century. By 2012 there were 12 known particles of matter – six called quarks and six called leptons. Quarks and leptons are now thought to be the smallest possible particles. These particles have been used in a model of matter to explain how the cosmos works. For many years scientists have been searching for one missing particle known as the Higgs boson. The discovery would prove the existence of the Higgs field, which gives particles mass.

On 4 July 2012 scientists at CERN, the European organisation for Nuclear Research, announced that they had discovered the Higgs boson. The particle was named after Professor Peter Higgs, a British scientist who, while working at Edinburgh University, was one of six scientists who had predicted its existence in the 1960s. The controversial term 'God particle' was used in the 1990s, emphasising its importance in explaining the origins of the universe.

Figure 1.2.1: *CERN*

The discovery of the Higgs boson has been described as the most important discovery in 50 years. It was discovered using an instrument called the Large Hadron Collider (LHC) in the biggest experiment in history. The LHC is a particle accelerator built underground near Geneva in Switzerland and has a circumference of 27km. It attempts to recreate the conditions that existed just after the 'Big Bang' (when the universe is thought to have been formed) by colliding beams of particles travelling near the speed of light. When the particles smash into each other at this speed, smaller, unstable, short-lived particles are formed and detected. To produce a Higgs particle, the LHC smashes protons together about a billion times every second, producing something like one Higgs particle every 10 billion collisions.

The type of activity being undertaken by scientists at CERN has already had an impact on society in areas such as health. Particle accelerators are routinely used in hospitals for cancer radiotherapy, and radioisotopes have been produced that are used in the detection and treatment of cancer (this is covered in more detail in Chapter 10: Nuclear chemistry). Cutting-edge scientific research requires state-of-the-art technologies. Technologies developed for use in the projects at CERN are now being used by industry and commerce.

GO! Nobel Prize winners

Professor Peter Higgs and François Englert were awarded the Nobel Prize in Physics in 2013. Professor Higgs had jointly predicted the existence of the Higgs boson in the 1960s while working at the University of Edinburgh. In 2012, the boson was detected in the Large Hadron Collider particle accelerator, which was built underground near Geneva in Switzerland.

Figure 1.2.2: *Professor Peter Higgs giving a press conference at the University of Edinburgh*

Figure 1.2.3: *Collision of particles in a Large Hadron Collider*

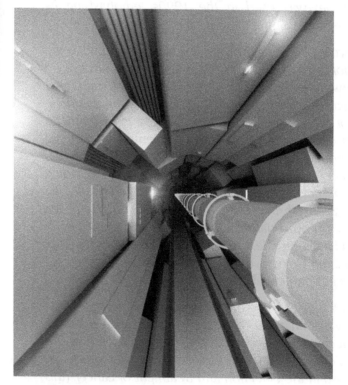

Figure 1.2.4: *The Large Hadron Collider (LHC) at CERN collides beams of particles, to create subatomic particles, including the Higgs boson*

Atomic structure

All substances are made up of atoms. Atoms contain particles called protons (p), neutrons (n) and electrons (e⁻). Figure 1.2.5 shows how these particles are arranged in atoms. The left-hand diagram gives the idea that atoms are not flat and electrons are moving. The right-hand diagram is called a 'target diagram' and shows the electrons arranged in shells, like layers in an onion.

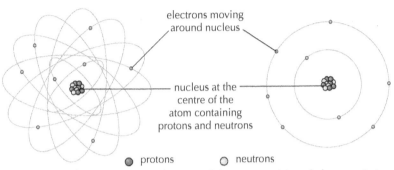

Figure 1.2.5: *The arrangement of protons (p), neutrons (n) and electrons (e⁻) in an atom*

Each particle has a charge and mass, as shown in Table 1.2.1.

Table 1.2.1: *Charge and mass of particles*

Particle	Charge	Mass
proton (p)	one positive (+)	1
electron (e–)	one negative (–)	almost zero
neutron (n)	no charge (0)	1

The number of protons in an atom identifies which element the atom belongs to. The number of protons is known as the **atomic number**. The elements are arranged in order of increasing atomic number in the periodic table.

In an atom the number of protons is the same as the number of electrons. The positive charge of the protons is balanced by the negative charge of the electrons so, overall, an atom is neutral (neutrons have no charge).

If the atomic number is known, the element can be identified and the number of protons and electrons in each atom of an element can be worked out.

An element has an atomic number of 11. Using the Periodic Table, the element can be identified as sodium. Each sodium atom must have 11 protons and 11 electrons.

Example 1.2.1

> 📖 **Word bank**
>
> • **Atom**
> An atom is the smallest, most basic unit of an element, containing particles called protons, neutrons and electrons.

> 📖 **Word bank**
>
> • **Atom**
> • **Atomic number**
> The atomic number of an element is given by the number of protons in an atom of that element. Atomic number is unique to each element.

If the name of the element is known, then the atomic number can be found in the Periodic Table, and this gives the number of protons and electrons.

Example 1.2.2

Fluorine has an atomic number of 9, so a fluorine atom has nine protons and nine electrons.

If the number of protons in an atom is known, the atomic number of the element will be the same, and the name of the element can be found from the Periodic Table.

Example 1.2.3

An atom has 18 protons. Its atomic number must be 18, and so the element is argon.

> ### 📖 Word bank
>
> • **Mass number**
> The mass number of an atom is the number of protons and neutrons added together.

The mass of an atom is due to the mass of the protons and neutrons. The mass of the protons added to the mass of the neutrons is known as the **mass number**. The electrons are so light that they don't affect the mass.

mass number = number of protons + number of neutrons

The number of neutrons can be calculated by subtracting the atomic number from the mass number.

number of neutrons = mass number – atomic number

Example 1.2.4

An atom of lithium has 4 neutrons. Calculate its mass number.

From the Periodic Table, the atomic number of lithium is 3. The atom must have 3 protons.

mass number = number of protons + number of neutrons

$$= \quad 3 \quad + \quad 4$$

$$= \quad 7$$

Periodic Table

> ### 📖 Word bank
>
> • **Valency**
> The valency of an atom of an element indicates how many bonds it can form with atoms of other elements.

Elements are substances made from the same atoms. The Periodic Table shows elements with similar chemical properties arranged in families. Elements in the same vertical group have the same valency and react in a similar way – they have similar chemical properties. Each element has its own symbol and atomic number. Metals are found on the left of the Periodic Table and non-metals to the right.

The electrons in an atom are arranged in an organised way. The electrons move around the nucleus in layers of space called energy levels (sometimes called shells), a bit like the layers in an onion. This can be shown in a target diagram.

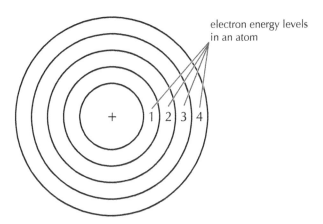

electron energy levels
in an atom

Figure 1.2.6: *Electron energy levels in an atom*

The number of electrons in the outer energy level of an atom is the same as the group number of the element in the Periodic Table. The electron arrangement for the elements are written in a way which shows how many electrons are in each energy level.

Example 1: Potassium (atomic number 19), in Group 1: 2,8,8,1

Example 2: Chlorine (atomic number 17), in Group 1: 2,8,7

 Hint

You only need to be able to write the electron arrangement of the first 20 elements. They are found in the SQA data booklet.

GO! Activity 1.2.1

1. (a) Copy and complete the following paragraph. You may wish to use the information in Table 1.2.1 to help you.

The atom is made up of (a)_____ (p), neutrons (b)_____ and (c)_____ (e^-). The protons and (d)_____ are found in the (e)_____ at the centre of the atom. The (f)_____ move around the nucleus. Protons have a (g)_____ charge and a mass of (h)_____. Neutrons have (i)_____ charge and a mass of 1. Electrons have a (j)_____ charge and a mass of almost (k)_____.

(b) Draw a 'target diagram' to show the arrangement of the particles that make up a hydrogen atom with 1 proton, 2 neutrons and 1 electron. Label your diagram clearly.

2. (a) The table below gives information about the atoms of three elements. Use a periodic table to help you complete entries **(a)–(o)**.

Element	Symbol	Atomic number	Protons	Electrons	Neutrons	Mass number
lithium	(a)	(b)	(c)	(d)	4	(e)
(f)	(g)	(h)	17	(i)	(j)	35
(k)	(l)	10	(m)	(n)	11	(o)

(b) Explain why atoms have a neutral charge, even although they are made up of positive and negative particles.

📖 Word bank

• **Nuclide notation**

A shorthand way of showing the mass number and atomic number of an atom along with the symbol of the element.

Nuclide notation

Chemists use shorthand and symbols to summarise reactions and represent elements and compounds. **Nuclide notation** is a way of showing the mass number and atomic number of an **atom** along with the symbol of the element. The nuclide notation for an atom of chlorine, atomic number 17, with a mass number of 35 can be used as an example:

🔵 Activity 1.2.2

An atom of aluminium has 14 neutrons and 13 protons.

(a) Write the nuclide notation for the atom.

(b) When an aluminium atom forms an ion it loses three electrons. Write the nuclide notation for the aluminium ion.

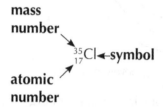

Decoding the chemical shorthand used for chlorine tells us:

atomic number = 17, so **no. of protons = 17**

no. of electrons = 17 (atoms are neutral)

mass number = 35, so **no. of neutrons = 35 – 17 = 18**

Nuclide notation can also be used for **ions**. The number of electrons is the only thing that changes when an atom forms an ion. This means the atomic number (number of protons) and the mass number (number of protons + neutrons) doesn't change. The nuclide notation for an atom and ion of both sodium and chlorine is shown below:

$$^{23}_{11}\text{Na}\begin{cases}23 - 11 = 12 \text{ neutrons} \\ 11 \text{ protons} \\ 11 \text{ electrons}\end{cases}$$

a sodium **atom**

$$^{23}_{11}\text{Na}^+\begin{cases}23 - 11 = 12 \text{ neutrons} \\ 11 \text{ protons} \quad (11+) \\ \mathbf{10 \text{ electrons}} \ (10-)\end{cases}\begin{matrix}\text{overall} \\ 1+\end{matrix}$$

a sodium **ion**

$$^{35}_{17}\text{Cl}\begin{cases}35 - 17 = 18 \text{ neutrons} \\ 17 \text{ protons} \\ 17 \text{ electrons}\end{cases}$$

an **atom** of chlorine

$$^{35}_{17}\text{Cl}^-\begin{cases}35 - 17 = 18 \text{ neutrons} \\ 17 \text{ protons} \ (17+) \\ \mathbf{18 \text{ electrons}} \ (18-)\end{cases}\begin{matrix}\text{overall} \\ 1-\end{matrix}$$

an **ion** of chlorine

Isotopes

📖 Word bank

• **Isotopes**

Atoms that have the same atomic number but different mass numbers.

Not all atoms of the same element have the same number of neutrons. Some atoms of chlorine, for example, have 18 neutrons while others have 20 neutrons. This means their mass numbers are different (35 and 37) but their atomic numbers are the same (17). Atoms which have the same atomic number but different mass numbers are known as **isotopes**. The nuclide notation for chlorine 37 is $^{37}_{17}\text{Cl}$.

Most elements exist as isotopes. Hydrogen has three isotopes: $^{1}_{1}\text{H}$ (sometimes called protium); $^{2}_{1}\text{H}$ (deuterium); $^{3}_{1}\text{H}$ (tritium). Deuterium and tritium are naturally occurring radioactive isotopes. Some radioactive isotopes are man-made and are widely used in medicine and industry (see Chapter 10: Nuclear chemistry). Table 1.2.2 shows the number of each subatomic particle in isotopes of chlorine and hydrogen:

Table 1.2.2

Element	Istotopes	Protons	Electrons	Neutrons
Chlorine	$^{37}_{17}Cl$	17	17	20
	$^{35}_{17}Cl$	17	17	18
Hydrogen	$^{1}_{1}H$	1	1	0
	$^{2}_{1}H$	1	1	1
	$^{3}_{1}H$	1	1	2

The chemical properties of an element are dependent on the electron arrangements of the atoms of the element. Since isotopes of an element have the same electron arrangement they will exhibit identical chemical behaviour.

It is the dream of scientists to provide a cheap source of power by fusing (joining) nuclei and capturing the huge amounts of energy produced during the process. Nuclear fusion needs a temperature greater than 100 million degrees centigrade and this is one of the main obstacles to be overcome. The reaction which is most promising involves the fusing of deuterium and tritium, two isotopes of hydrogen. The chemical product is also important as it is helium, which is harmless and will have no environmental impact. The latest experimental reactor ITER (International Thermonuclear Experimental Reactor) is being built in France. The aim is for the reactor to produce at least ten times as much energy as is put in to bring about the fusion process.

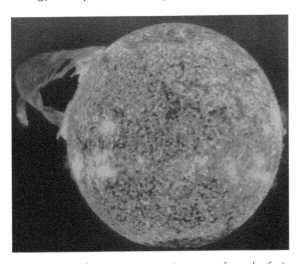

Figure 1.2.7: *The Sun generates its energy from the fusion reaction between the hydrogen isotopes deuterium and tritium*

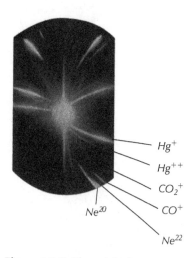

Hg$^+$

Hg^{++}

CO$_2^+$

Ne^{20}

CO$^+$

Ne^{22}

Figure 1.2.8: *The original photographic plate from Thomson's experiment showing the markings for the two isotopes of neon: neon-20 and neon-22*

STEP BACK IN TIME: THE DISCOVERY OF STABLE ISOTOPES

The existence of isotopes had been proposed to explain aspects of radioactivity by Frederick Soddy when working at Glasgow University at the beginning of the twentieth century. The first evidence for isotopes of a stable (non-radioactive) element was found by J.J. Thomson in 1913 as part of experimentation which led to the discovery of positive ions. Thomson channelled streams of neon ions through a magnetic and electric field and measured their deflection by placing a photographic plate in their path. Each stream created a glowing patch on the plate at the point it struck. Thomson observed two separate patches of light on the photographic plate and eventually concluded that some of the atoms in the neon gas were of higher mass than the rest.

His research assistant, F.W. Aston, later discovered different stable isotopes for a number of elements using an instrument called a mass spectrograph. In 1919 Aston was able to show that neon had two isotopes, 20 and 22, and that neither was equal to the known Relative Atomic Mass of neon gas (20.2). Aston similarly showed that the relative atomic mass of chlorine (35·45) is a weighted average of the two isotopes Cl-35 and Cl-37, which takes into account the percentages of each of the two isotopes.

J.J. Thomson's separation of neon isotopes by their mass was the first example of mass spectrometry, which was then improved and developed into a general method by F.W. Aston and A.J. Dempster.

SPOTLIGHT ON FORENSIC SCIENCE

Identifying the quantity of specific isotopes in a material can be used to work out where it has come from. A recent development in forensic science is the isotopic analysis of hair strands. Hair grows at an average rate of 1cm a month. One of the main factors affecting hair growth is water intake. The amount of stable isotopes in drinking water depends on the type of earth it has come through, which varies throughout the world. These isotope differences are then

Make the Link

The reliance of life on water is covered in biology.

biologically 'set' in our hair as it grows and it has therefore become possible to identify where a person has recently been in the world by the analysis of hair strands. For example, it could be possible to identify whether a terrorist suspect had recently been to a particular location from hair analysis. This type of analysis has also been used to identify human remains. Hair samples are easy to obtain and this form of analysis is becoming very popular in cases where DNA or other traditional means are bringing no answers.

Isotope analysis can be used to work out whether two or more samples of explosives have come from the same place. Most high explosives contain carbon, hydrogen, nitrogen and oxygen atoms and comparing the relative amounts of isotopes can indicate they came from the same location.

Isotopic analysis has also been used to help identify drug trafficking routes. The isotopic quantities in morphine grown from poppies in south-east Asia are different to poppies grown in south-west Asia. Similarly, investigators have been able tell if cocaine has come from Bolivia or Colombia using isotopic analysis.

Other forensic uses of isotopes include:

- Analysing cellulose used in the manufacture of banknotes by comparing the amounts of hydrogen and oxygen isotopes to work out where counterfeit money came from.

- Analysing various materials such as tyres, cigarette filters, packaging tapes, sawdust, human blood, hair, paint, varnish and micro-organisms for evidence in criminal cases.

Relative Atomic Mass (RAM)

The total mass of an atom comes from the mass of its neutrons and protons. The electrons are so light that they don't add significantly to the overall mass. Most elements, however, have two or more isotopes so an average is taken of the mass of all the isotopes. This average mass is called the **Relative Atomic Mass (RAM)**.

Relative atomic masses have no units because they are measured relative to each other and are seldom whole numbers but are often rounded to the nearest 0·5, as shown in the table of selected RAM values.

> ### 📖 Word bank
> - **Relative Atomic Mass (RAM)**
> Average of the masses of the isotopes of an element.

Table 1.2.3: *Table of selected RAM values*

Element	Symbol	Relative atomic mass	Element	Symbol	Relative atomic mass
aluminium	Al	27	magnesium	Mg	24·5
argon	Ar	40	mercury	Hg	200·5
bromine	Br	80	neon	Ne	20
calcium	Ca	40	nickel	Ni	58·5
carbon	C	12	nitrogen	N	14
chlorine	Cl	35·5	oxygen	O	16
copper	Cu	63·5	phosphorus	P	31
fluorine	F	19	platinum	Pt	195
gold	Au	197	potassium	K	39
helium	He	4	silicon	Si	28
hydrogen	H	1	silver	Ag	108
iodine	I	127	sodium	Na	23
iron	Fe	56	sulfur	S	32
lead	Pb	207	tin	Sn	118·5
lithium	Li	7	zinc	Zn	65·5

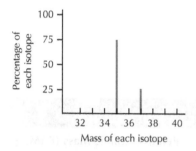

Figure 1.2.9: *Graph showing percentage of the two isotopes of chlorine*

The relative atomic mass of chlorine is taken as 35·5. The two isotopes of chlorine are ^{35}Cl and ^{37}Cl. If the average mass is 35·5 then there must be more isotopes with mass number 35 than 37. We get this information from an instrument called a mass spectrometer. It tells us: how many isotopes there are; the mass of each isotope and the percentage of each isotope. The information is often given in a graph like the one shown for chlorine. From the graph we can see that:

- Chlorine has two isotopes – corresponding to the two peaks.
- 75% of atoms have a mass of 35 and 25% have a mass of 37.

This information can be used to calculate the relative atomic mass of chlorine:

RAM = [(75 × 35) + (25 × 37)] / 100

RAM = 35·5

GO! Activity 1.2.3

1. The table gives information about the three naturally occurring isotopes of silicon.

Isotope	Percentage abundance
^{28}Si	92
^{29}Si	5
^{30}Si	3

Calculate the relative atomic mass of silicon.

2. The table shows the RAM of some of the Group 1 elements and their melting points.

Element	RAM	Melting point/°C
Li	7·0	180
Na	23·0	98
K	39·0	64
Rb	85·5	39

(a) Draw a line graph of melting point against RAM.
(b) From the graph, predict the melting point of caesium (RAM 133·0).

3. A **mass spectrometer** separates the **isotopes** of an element according to their **mass number**. The information can be used to calculate the **relative atomic mass** of an element. The graph obtained for the element magnesium, is shown below.

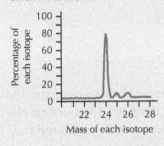

(a) Explain the meaning of the terms shown in bold.
(b) Use the graph to answer the following questions:
   (i)   How many isotopes does magnesium have?
   (ii) Which isotope is there most of?
   (iii) Explain why the relative atomic mass of magnesium is 24·5.

Covalent bonding and the shapes of molecules

The noble gases (Group 0) are very stable elements – they are extremely unreactive. They are so stable that they exist as single unbonded atoms – they are monatomic. Table 1.2.4 shows the electron arrangements for some of the noble gases. For each element the outer energy level is full. This is what makes them unreactive.

Table 1.2.4: *The electron arrangements for some of the noble gases*

Element	Electron arrangement	
He	2	first energy level full
Ne	2.8	second energy level full
Ar	2.8.8	third energy level full

The atoms of elements in other groups do not have full outer energy levels. These atoms are unstable. Atoms of non-metal elements achieve a stable electron arrangement by sharing electrons and forming a covalent bond. Figure 1.2.10 shows two fluorine atoms coming together to form a molecule – only the outer energy level electrons are shown.

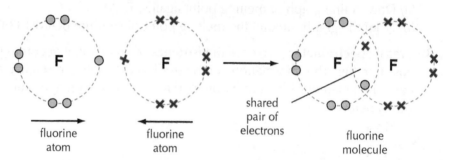

Figure 1.2.10: *Two fluorine atoms come together to form a molecule*

This type of diagram shows electrons as dots and crosses and is often referred to as a **'dot and cross' diagram**. The fluorine atoms are held together by a single covalent bond – a shared pair of electrons.

At first sight it might seem unlikely that atoms would be attracted because the shells of electrons around the atoms would repel each other, but the positive nucleus of each atom attracts not only its own electrons but also electrons from the other atom. This is shown diagrammatically in Figure 1.2.11.

Count the electrons in the outer energy level of each atom in the fluorine molecule in Figure 1.2.10 – each has seven electrons of its own and a share in an eighth. Eight electrons is a stable arrangement – each fluorine atom in the molecule now has the

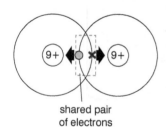

Figure 1.2.11: *The nucleus of one atom attracts the electrons of the other*

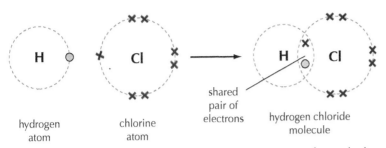

Figure 1.2.12: *A hydrogen atom and a chlorine atom join to form a hydrogen chloride molecule*

same electron arrangement as the noble gas neon. It is, of course, not neon because each atom still has nine protons, whereas neon atoms have ten protons.

The diagram in Figure 1.2.12 shows a hydrogen atom and a chlorine atom joining to form a hydrogen chloride molecule – only the outer electrons are shown for the chlorine atom.

Count the number of electrons in the outer energy level of each atom in the molecule. Chlorine now has eight electrons – a stable arrangement. Hydrogen now has two electrons, which at first glance is not a stable arrangement. However it is the same arrangement as the noble gas helium – two electrons in its first (only) energy level is a stable arrangement.

Fluorine and hydrogen chloride exist as diatomic molecules – fluorine is an element and hydrogen chloride a compound. They have a **linear** shape, i.e. their atoms are in a line. The covalent bond can be represented as –. A fluorine molecule can be written as F–F and hydrogen chloride as H–Cl. This representation shows both the covalent bond and the shape of the molecule.

Atoms of oxygen and hydrogen combine by **sharing pairs of electrons** to form water (H_2O). The oxygen atom has six outer electrons so needs two more. Oxygen forms two single covalent bonds to hydrogen atoms:

> **📖 Word bank**
>
> • **Dot and cross diagram**
> Dots and crosses are used to represent electrons in diagrams showing how the atoms bond.

> **📖 Word bank**
>
> • **Shared pair of electrons**
> When two electrons from two atoms are shared a covalent bond is formed.

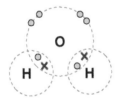

Figure 1.2.13: *Shared pairs of electrons in a water molecule*

The molecular shape for water is described as **angular**:

Figure 1.2.14: *A water molecule is angular*

Atoms of nitrogen and hydrogen combine by sharing pairs of electrons to form ammonia (NH_3). The nitrogen atom has five outer electrons so needs three more. Nitrogen forms three single covalent bonds to hydrogen atoms:

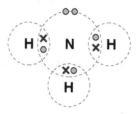

Figure 1.2.15: *Shared pairs of electrons in an ammonia molecule*

The molecular shape for ammonia is **trigonal pyramidal**.

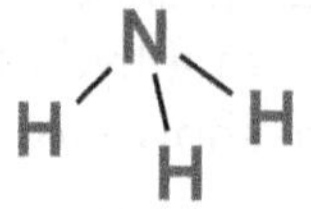

Figure 1.2.16: *An ammonia molecule has a trigonal pyramidal shape*

Atoms of carbon and hydrogen combine by sharing pairs of electrons to form methane (CH_4). The carbon atom has four outer electrons so needs four more. Carbon forms four single covalent bonds to hydrogen atoms:

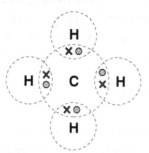

Figure 1.2.17: *Shared pairs of electrons in a methane molecule*

The molecular shape for methane is **tetrahedral**.

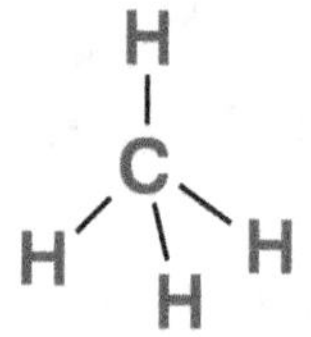

Figure 1.2.18: *A methane molecule is tetrahedral*

The table gives a general rule about the number of atoms in a molecule with single bonds and the shape of the molecule:

Table 1.2.5

Number of atoms in molecule	Shape of molecule
two	linear
three	angular
four	trigonal pyramidal
five	tetrahedral

GO! Activity 1.2.4: Paired activity

You can discuss these activities with a partner. Make sure you check your answers.

1. (a) Draw a dot and cross diagram to show how atoms of sulfur and hydrogen form covalent bonds with each other.
   (b) Using your diagram from (a) above write the molecular formula for the compound and name it.
   (c) What shape is the molecule likely to be?
   (d) The shape of the molecule in (c) is the same as the shape of a molecule of water. Why might you have been able to predict that the molecules would have had the same shape given the position of oxygen and sulfur in the Periodic Table?

2. (a) Draw a dot and cross diagram to show how atoms of silicon and hydrogen form covalent bonds with each other.
   (b) Using your diagram from (a) above write the molecular formula for the compound and name it.
   (c) What shape is the molecule likely to be?
   (d) The shape of the molecule in (c) is the same as the shape of a molecule of methane. Why might you have been able to predict that the molecules would have had the same shape given the position of silicon and carbon in the Periodic Table?

3. Collect a box of molecular models and make models of fluorine, hydrogen chloride, water, ammonia and methane. Note the shape of each molecule and compare them to the diagrams in the section above.

   The 'atoms' are usually coloured as follows:

   hydrogen – white

   chlorine and fluorine – green

   oxygen – red

   nitrogen – blue

   carbon – black.

It is possible for the atoms of non-metal elements to join together in such a way as to form double and even triple covalent bonds. In an oxygen molecule (O_2) both oxygen atoms have six outer electrons so both need two more electrons to form a stable arrangement of electrons (eight). They do this by sharing two pairs of electrons forming a double covalent bond:

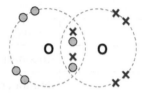

Figure 1.2.19: *Two shared pairs of electrons in an oxygen molecule*

This can also be shown as:

$$O = O$$

Figure 1.2.20: *The double bond in an oxygen molecule*

In a nitrogen molecule (N_2) both nitrogen atoms have five outer electrons so both need three more electrons to form a stable arrangement of electrons. They do this by sharing three pairs of electrons forming a triple covalent bond:

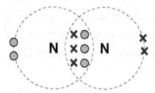

Figure 1.2.21: *Three shared pairs of electrons in a nitrogen molecule*

This can be shown as:

$$N \equiv N$$

Figure 1.2.22: *The triple bond in a nitrogen molecule*

This triple bond within the nitrogen molecule is the reason nitrogen is so unreactive and needs a lot of energy to break the bond. This explains why nitrogen and oxygen, which make up almost all of the air we breathe, don't react unless there is a lightning storm. It makes the production of ammonia and nitric acid much harder than it should be theoretically (see Chapter 9: Fertilisers).

In the compound carbon dioxide (CO_2) carbon has four outer electrons so needs four more. It forms double covalent bonds with two oxygen atoms so that all the atoms now have a full outer shell of electrons:

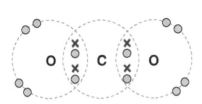

Figure 1.2.23: *The shared pairs of electrons in a carbon dioxide sample*

This can also be shown as:

$$O = C = O$$

Figure 1.2.24: *The carbon dioxide molecule is linear*

Note that the shape of the carbon dioxide molecule is linear, because of the double bonds, and not bent as might be expected.

🔦 SPOTLIGHT ON THE AMAZING ELECTRON

*Despite its tiny mass, electrons are responsible for the chemical reactions of elements and compounds. They are also the reason we have **lasers** and **electron microscopes**.*

***Laser** stands for 'light amplification by stimulated emission of radiation'. All atoms absorb energy, which results in electrons moving to a higher energy level. When an electron drops back down to a lower energy level it releases its energy as a photon – a small bundle of light. A laser is a device which controls the way that energised atoms release photons with the same frequency. The energy produced in a laser is concentrated into a very tight beam, unlike the beam of a torch where the light comes out in all directions. The beam in a laser moves back and forward between two mirrors exciting more atoms and intensifying the energy in the beam. One of the mirrors allows some of the light out – this is the laser light. Though there are many different types of laser, they all work in a similar way.*

Lasers were first invented in the 1960s but became part of daily life with the introduction of the supermarket barcode scanner in 1974. The laserdisc player, produced in 1978, was the first successful consumer product to include a laser, and was followed by the hugely popular compact disc (CD) player, in 1982, and then by laser printers.

The use of lasers is now widespread and includes military, medical and scientific applications as well as consumer goods. Laser eye surgery is very common nowadays. Sensitive eye tissue can be removed with a laser without causing any heat damage to the neighbouring material.

The National Ignition Facility (NIF) sited in Livermore, California, is the world's largest and most energetic laser. The aim of the NIF is to achieve energy production by

Figure 1.2.25: *Scanning barcodes in the supermarket*

💥 Make the Link

Energy is a subject covered in depth in physics.

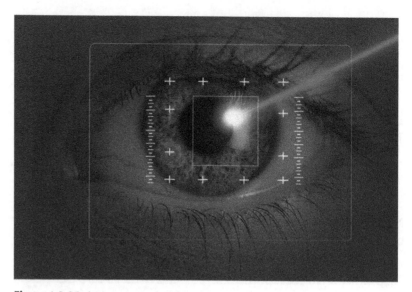

Figure 1.2.26: *Laser eye surgery is now a common procedure*

nuclear fusion, in the laboratory, for the first time. It has been described as creating a mini-star on Earth. The NIF will focus the intense energy of 192 giant laser beams on a target filled with hydrogen fuel, fusing the hydrogen atoms' nuclei and releasing many times more energy than it takes to start the fusion reaction. NIF's laser beams can produce more than 60 times the energy of any previous laser system.

Electron microscopes *allow us to see images of very small objects that can't be seen with ordinary microscopes. Electron microscopy uses the charge on the electrons around atoms to make images of separate atoms and molecules. This has been particularly important in areas like medical research where it is now possible to see viruses and other organisms that cause disease.*

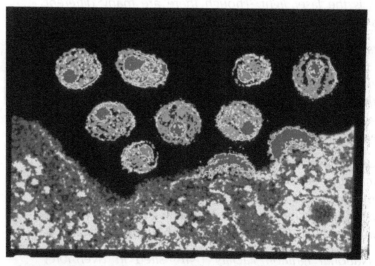

Figure 1.2.27: *Electron micrograph of AIDS viruses around a white blood cell*

In recent years the scanning probe microscope has been developed, which involves probing the surface of a substance with a fine conducting tip. This has given rise to a new branch of science called **nanotechnology,** *which involves manipulating materials on an atomic or molecular scale with a view to building microscopic machines. A nanometer (nm) is one-billionth of a meter, which is a hundred-thousandth of the width of a human hair.*

SPOTLIGHT ON NANOTECHNOLOGY

Scientists have found two nano-size structures of particular interest: nanowires and carbon nanotubes. Nanowires are wires with a very small diameter, sometimes as small as one nanometer. Scientists hope to use them to build tiny transistors for computer chips and other electronic devices. A carbon nanotube is a nano-size cylinder of carbon atoms. With the right arrangement of atoms, it is possible to create a carbon nanotube that is hundreds of times stronger than steel, but six times lighter. Engineers plan to make building material out of carbon nanotubes, particularly for things like cars and airplanes. Lighter vehicles would mean better fuel efficiency, and the added strength means increased passenger safety.

Nanotechnology is making a big impact on the tennis world. In 2002, the tennis racket company Babolat made a racket out of

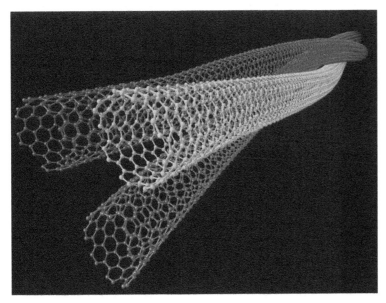

Figure 1.2.28: *Diagram showing the structure of a carbon nanotube*

carbon nanotube-infused graphite, meaning the racket was very light, yet many times stronger than steel. The tennis ball manufacturer Wilson introduced the Double Core tennis ball. These balls have a coating of clay nanoparticles on the inner core. The clay acts as a sealant, making it very difficult for air to escape from the ball.

There are a number of other products on the market that use nanotechnology. Many sunscreens contain nanoparticles of zinc oxide or titanium oxide. Older sunscreen formulas use larger particles, which is what gives most sunscreens their whitish colour. Smaller particles are less visible, meaning that when you rub the sunscreen into your skin, it doesn't give you a whitish tinge. Some antibacterial bandages use nanoparticles of silver to smother harmful cells, killing them. Other uses include self-cleaning glass (see Chapter 11: Chemical analysis), improving scratch-resistant surfaces, coating clothing to make it stain resistant and give better UV protection.

Eric Drexler, the person who introduced the word 'nanotechnology', has issued a warning about its use saying that there may be a danger of self-replicating nanorobots going wrong, duplicating themselves a trillion times over and consuming the entire world as they pull carbon from the environment to build more of themselves. It's called the 'grey goo' scenario, where a synthetic nano-size machine replaces all organic material. Another scenario involves nanodevices made of organic material wiping out the Earth – the 'green goo' scenario. This may seem like science fiction but so did nanotechnology not that long ago.

Structure and properties of covalent substances

Covalent substances can form either **individual molecules** or **giant networks**. All of the substances mentioned in the previous section Covalent bonding and the shapes of molecules are **covalent molecular** and exist as either gases, e.g. methane, or liquids, e.g. water, at room temperature. This is because, although the covalent bonds holding the atoms together are relatively strong, the forces between the molecules are weak and it doesn't take a lot of energy to separate the molecules. It is these weak forces between molecules that have to be broken when a molecular substance melts or boils. This results in molecular substances generally having low melting and boiling points. Carbon dioxide (CO_2) is a gas at room temperature because there is enough energy to overcome the weak forces of attraction between the molecules:

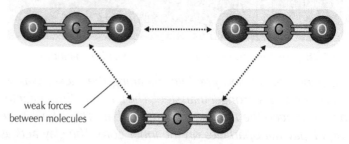

weak forces between molecules

Figure 1.2.29

Some covalent molecular substances are solid at room temperature because of their molecular size – the larger the molecule the bigger the forces of attraction between the molecules and the more energy needed to separate them. Phosphorous molecules consist of four atoms (P_4) and sulfur molecules consist of eight (S_8). Table 1.2.6 compares the melting and boiling points of some non-metal elements.

Table 1.2.6: *Melting and boiling points of some non-metallic elements*

Element	Molecular formula	Melting point/°C	Boiling point/°C	State at room temperature
nitrogen	N_2	−210	−196	gas
oxygen	O_2	−218	−183	gas
bromine	Br_2	−7	59	liquid
phosphorus	P_4	44	280	solid
sulfur	S_8	113	445	solid

Some covalently bonded substances exist as solids with very high melting and boiling points. They can't exist as individual molecules otherwise their melting and boiling points would be much lower. They are made up of giant structures called **networks** in which each atom is covalently bonded to another in a giant three-dimensional structure. Covalent bonds are relatively strong and it takes a lot of energy to break the bonds. Silicon dioxide (SiO_2) is a strong, solid covalent network substance with a melting point = 1610°C and a boiling point = 2230°C. Diamond is made up of pure carbon in which each of the four outer electrons in each carbon atom are covalently bonded to a different carbon atom resulting in a covalent network structure. Diamond is the hardest known substance.

Some covalent substances like sucrose (sugar) are soluble in water but most are not. Many are however soluble in covalent solvents. You may have cleaned paint brushes in white spirit because the paint won't dissolve in water. White spirit is the most widely used solvent in the paint industry. Nail varnish, a covalent solid, is dissolved in a covalent liquid called acetone. Nail varnish remover also contains covalent solvents. Many paints and varnishes have distinctive odours because their solvents have low boiling points and evaporate quickly so the paint or varnish dries quickly. Substances which contain covalent solvents have hazard warnings on their label as they can be harmful if inhaled and tend to be flammable.

Covalent substances are non-conductors of electricity because they do not have charged particles which are free to move.

● Carbon atom

○ Silicon atom

○ Oxygen atom

Figure 1.2.30: *Diamond and silicon dioxide are covalent network structures*

Figure 1.2.31: *Many common products contain covalent solvents*

Putting covalent network properties to use

Silicon dioxide, also known as quartz or silica, is the main constituent of sand used in the manufacture of glass. The most common glass made, soda lime glass, contains about 72% silica. Some 90% of all glass made is soda-lime glass and it is used for windows, containers, light bulbs and drinking glasses. Specialist glass can be made by adding other chemicals, such as boron oxide, which results in borosilicate glass that is more heat resistant than soda lime glass. One well-known trade name for this sort of glass is Pyrex. Quartz crystals are used as a semi-precious jewel and as a component in optical equipment.

Diamond is the hardest substance known and this property makes it very useful as an abrasive and as a cutting tool in industry and surgery.

In its purest form diamond is completely transparent and has no colour, which makes it highly valued as a precious stone.

Silicon carbide (SiC) has a diamond-like structure and has similar properties to diamond. Although it occurs in nature as the extremely rare mineral moissanite, most silicon carbide is synthetic. Its hardness and high melting point makes it useful as an abrasive, e.g. in certain sandpapers and for cutting discs. Fixing grains of silicon carbide to a variety of materials gives the materials non-slip properties, for example, the grip tape on skateboards. Grip tape is applied to the top surface of a board to allow the rider's feet to grip the surface and help the skater stay on the board while doing tricks. Synthetic moissanite has become popular as a diamond substitute in jewellery.

Figure 1.2.32: *A diamond-tipped oil drill part*

Figure 1.2.33: *Artificial diamond (silicon carbide) looks like the real thing*

◾◾€ SPOTLIGHT ON MATERIALS FOR THE FUTURE

Materials breakthrough wins Nobel

This was one of the headlines in 2010 when two scientists, Andre Geim and Konstantin Novoselov, both at Manchester University, won the prize for research on **graphene**.

Figure 1.2.34: *Silicon carbide grip tape on the surface of a skateboard helps the rider keep a grip on the board*

Figure 1.2.35: *Professor Andre Geim with a model of graphene, the so-called 'miracle material'*

Graphene is a flat sheet of carbon atoms covalently bonded, just one atom thick. It is almost completely transparent, 100 times stronger than steel, more conductive than copper and more flexible than rubber. Its unique properties mean it could have a wide range of practical uses. One of the ways graphene can be obtained is by heating silicon carbide (SiC) to high temperatures (over 1100°C) under low pressure.

In 2012 the UK government, realising the potential of graphene, allocated £50m to graphene research. The European Commission chose graphene as Europe's first ten-year, 1000 million euro Future Energy Technologies flagship in 2013, with the aim of taking graphene development out of the laboratory and into society.

There are high hopes for using graphene in the electronics industry – perhaps one day replacing silicon. It also has potential for use in the energy field in the form of advanced lightweight batteries.

Flexible electronic screens may be the first commercial use of graphene, one idea being 'e-paper' (electronic paper) which is designed to have the appearance of ordinary paper.

Graphene is so thin that a 'paint' could possibly act as a rust protector or an 'electronic ink'. A small amount of graphene mixed into plastics makes them conduct electricity.

Graphene's purity and large surface area could also make it suitable for medical uses, from aiding drug delivery to building new human tissue. However, some experts say that once again the cost and technical difficulties mean this will not happen before 2030.

It is the same with most of the proposed uses of graphene – it has to make financial sense to make it worth switching from existing materials to graphene.

One key indicator of the potential success of any new product is how many patents (right to use) a country and companies take out. At the start of 2013 there were more than 7000 patents on graphene, with the largest number – more than 2000 – held by China. The South Korean company Samsung holds more than 400 patents – they seem convinced that the 'miracle material' will be a success.

'Frozen smoke' is one of the nicknames given to a material made from silicon dioxide (silica), called **silica aerogel**. Despite the name, aerogel is a rigid, solid, dry material and does not resemble a gel in its physical properties. The name

Figure 1.2.36: *A block of silica aerogel*

comes from the fact that it is made from a gel. Silica (SiO_2) aerogel is the most common type of aerogel. It is made by extracting the liquid from the gel in a way that preserves 90–99% of the gel framework's original volume.

The silica chains in the structure encircle air-filled pores. This makes aerogels incredibly light (low density) and gives them tremendous heat insulating properties.

Silica aerogels were used as insulation on the Mars Exploration Rovers, Spirit and Opportunity. The temperature on Mars ranges from −140°C to +20°C, which would cause severe damage to sensitive electronic parts if they weren't well insulated. The Stardust space mission used silica aerogel to collect dust particles from the comet Wild 2. Layers of silica aerogel were able to safely and softly cushion the impact of the particles travelling at around 13 600 miles an hour. The particles were embedded in the aerogel and analysed back on Earth.

Figure 1.2.37: *The silica aerogel particle collector from the Stardust space mission*

The use of silica aerogel as insulation in transparent window glass is being developed. One of the problems being encountered is making aerogels the size of a large window pane.

Silica aerogel is water repellant but very good at absorbing oil. As a result, special aerogel powder and blankets have been developed for cleaning up oil spillages.

As with many other modern materials, the commercial success of aerogel depends to a great extent on whether it can be produced cheaply enough to replace existing materials.

Hint

You can follow the development of graphene and silica aerogel by typing their names into an internet search engine.

⊙ Activity 1.2.5: Paired activity

You can discuss these activities with a partner. Make sure you check your answers.

1. Complete the summary for covalent bonding in your notebook – you may wish to use the word bank to help you:

Covalent substances exist as either (a)_____ molecules or giant (b)_____.
Molecular substances are mainly gases or (c)_____. This is because although the
(d)_____ bonds which hold the atoms together are relatively (e)_____ the forces
between the molecules are (f)_____. It does not take a lot of (g)_____ to (h)_____
the molecules. This results in molecular substances having (i)_____ melting and boiling
points. Covalent network substances have (j)_____ melting and boiling points. This is
because each atom in the network is (k)_____ bonded to other atoms resulting in a
very strong (l)_____ structure which needs a lot of energy to break the (m)_____.

Word bank
three-dimensional, bonds, covalent, covalently, energy, high, individual, liquid, low,
networks, separate, strong, weak

2. Carbon and silicon are in the same group of the Periodic Table and it could be predicted
that their oxides would have similar properties. Explain then why carbon dioxide is a gas
with low melting and boiling points while silicon dioxide is a hard solid with high
melting and boiling points.

Structure and properties of ionic compounds

Ions and ionic bonding

📖 Word bank

• **Ion**
An ion is a charged atom. It can be positive or negative.

When metal atoms react with non-metal atoms, the metal atom transfers electrons to the non-metal atom.

When a metal atom loses an electron it becomes positively charged. The non-metal atom that gains the electron becomes negatively charged. Charged atoms are called **ions**.

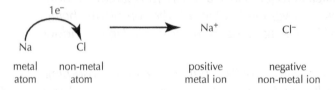

| metal atom | non-metal atom | positive metal ion | negative non-metal ion |

📖 Word bank

• **Ionic bond**
An ionic bond is formed when positive and negative ions attract each other.

Each ion now has the electron arrangement of a noble gas, which makes it very stable.

The attraction between the positive and negative ions results in the formation of an **ionic bond**.

In ionic compounds the oppositely charged positive (metal) and negative (non-metal) ions form a giant structure known as an **ionic lattice**. Sodium chloride (NaCl) is a good example of an ionic lattice.

Figure 1.2.38: *Sodium chloride crystals are cubic shaped*

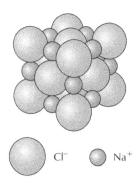

Cl⁻   Na⁺

Figure 1.2.39: *The ions in sodium chloride form a three-dimensional lattice*

There are strong **electrostatic forces** of attraction between the oppositely charged ions. This results in each ion being surrounded by ions carrying the opposite charge. If you examine the sodium chloride structure you will see that each sodium ion is surrounded by six chloride ions and each chloride ion is surrounded by six sodium ions. Eventually a giant three-dimensional structure (ionic lattice) is produced.

> ### 📖 Word bank
>
> • **Electrostatic forces**
> Attraction between oppositely charged ions.

Ionic compounds have many properties in common. These properties include:

- **High melting and boiling points** – it takes a lot of energy to break the ions apart. Table 1.2.7 shows the melting and boiling points of some Group 1 metal chlorides.

Table 1.2.7: *Melting and boiling points of some ionic compounds*

Compound	Formula	mp/°C	bp/°C
lithium chloride	LiCl	605	1350
sodium chloride	NaCl	801	1413
potassium chloride	KCl	770	1680

- **Solubility in water** – the regular pattern of the ions in the lattice is broken down and the ions are surrounded by water molecules.

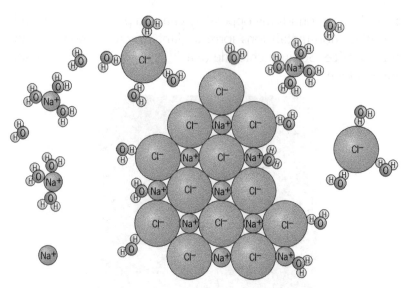

Figure 1.2.40: *The sodium chloride crystal lattice breaks down when it dissolves in water*

The electrostatic attraction between the ions is replaced by attractive forces between the ions and the water molecules. Ions in solution are often referred to as aqueous ions. When sodium chloride dissolves, aqueous sodium ions and aqueous chloride ions are formed:

$$NaCl(s) \rightarrow Na^+(aq) + Cl^-(aq)$$

The symbols $Na^+(aq)$ and $Cl^-(aq)$ represent the respective ions surrounded by water molecules.

Table 1.2.8: *Solubility of some ionic compounds in cold water: vs = very soluble s = soluble i = insoluble*

	bromide	carbonate	chloride	iodide	nitrate	phosphate	sulfate
ammonium	vs	vs	vs	vs	vs	vs	vs
barium	vs	i	vs	vs	vs	i	i
calcium	vs	i	vs	vs	vs	i	s
lithium	vs	vs	vs	vs	vs	i	vs
potassium	vs	vs	vs	vs	vs	vs	vs
sodium	vs	vs	vs	vs	vs	vs	vs

Sodium chloride and potassium chloride make up part of the mixture of chemicals in rehydration mixtures used to treat dehydration. Dehydration as a result of fluid loss caused by diarrhoea has been responsible for many deaths in some developing countries. The chemicals in rehydration mixtures are very soluble in water and are quickly absorbed into the body replacing essential salts. This cheap and effective treatment has saved millions of lives, especially children.

- **Electrical non-conductors as solids** – the ions are not free to move in the lattice.

- **Electrical conductors when in solution or a molten liquid** – the ions are free to move and act as charge carriers.

  When a d.c. supply is used, the ions move to the oppositely charged electrode, where they reform the neutral atom and can be identified.

Figure 1.2.41: *Rehydration powders contain soluble ionic compounds and their use has saved millions of lives*

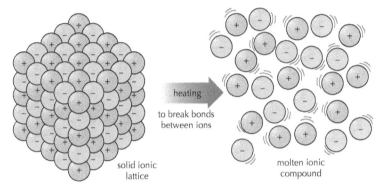

solid ionic lattice

heating to break bonds between ions

molten ionic compound

Figure 1.2.42: *Ionic compounds conduct when melted because the ions are free to move*

Ionic or covalent?

It is sometimes difficult to tell just by looking if a substance is ionic or covalent. Some simple observations can be made and tests carried out:

Colour
The colour of the substance can often indicate the type of bonding – many **ionic** substances are highly coloured.

State
If a substance is a **gas or liquid** at room temperature then the substance will be **covalently** bonded and exist as individual molecules. **Solids** are more difficult to distinguish from each other. If a solid can be **easily melted** then it will be **covalent molecular**. Wax (a hydrocarbon) melts quickly in boiling water indicating that it is covalent molecular. The liquid wax floats on the surface of the water and doesn't dissolve, which indicates it is covalent not ionic.

Figure 1.2.43: *The blue colour in copper sulfate crystals and the red in the gemstone ruby indicate the presence of ions*

Conductivity
Liquid wax doesn't conduct electricity, verifying that it is not ionic. The same test carried out on an ionic compound like sodium chloride, when it is in the liquid state, would have the opposite result, indicating ionic bonding.

GO! Activity 1.2.6: Paired activity

Carry out this activity with a partner and discuss your answers.

1. The table shows the properties of three substances, A, B and C.

Substance	Melting point	Electrical conductivity solid	liquid	solution
A	low	x	x	x
B	high	x	x	x
C	high	x	o	o

x = non-conductor   o = conductor

(a) Using the information in the table, state whether the bonding in the substances A, B and C is ionic, covalent molecular or covalent network.
(b) Justify each of your choices in (a).
(c) Explain why C is a conductor of electricity when it is a liquid or in solution but is a non-conductor when it is solid.

2. You are given a number of compounds. By observation of their appearance only, how could you tell which **might** be ionic and which **might** be covalent?

3. A student was asked to select some ionic compounds from Tables 1.2.7 and 1.2.8 to test their conductivity as solids, liquids and in solution.

His choices are listed below.

Testing conductivity in:

(a) The **solid** state: any of the compounds could be tested.
(b) The **liquid** state: lithium chloride rather than sodium chloride or potassium chloride.
(c) In **solution**: calcium carbonate.

Comment on his choice in each case.

Learning checklist

In this chapter you have learned:

- Atoms are made of protons (p), neutrons (n) and electrons (e^-).

- Protons have a mass of 1 and a positive charge.

- Electrons have practically no mass and a negative charge.

- Neutrons have a mass of 1 and no charge.

- Protons and neutrons are found in the nucleus and the electrons move around the nucleus.

- How to draw a 'target diagram' to show how protons, neutrons and electrons are arranged in an atom.

- How to work out the atomic number of an element using: atomic number = number of protons.

- How to work out the mass number of an atom using: mass number = number of protons + number of neutrons.

- An atom has a neutral charge because there are the same number of protons and electrons in an atom and their charges balance each other.

- A molecule is a small grouping of non-metal atoms bonded together.

- How to use nuclide notation for an atom and ion.

- Most elements have isotopes – not all atoms of an element have the same number of neutrons.

- Relative atomic mass (RAM) is the average mass of the isotopes and can be calculated using information from a mass spectrometer.

- Atoms are held together in a covalent bond because of the attraction of the nucleus of one atom to the outer electrons of another.

- Dot and cross diagrams can be drawn to show how atoms share a pair of electrons to form a single covalent bond.

- Multiple bonds can be formed between the atoms in some covalent molecules.

- The shapes of covalent molecules can be drawn.

- Covalent molecular substances have low melting and boiling points because the forces of attraction between them are very weak and not a lot of energy is needed to separate the molecules.

- A covalent network is a giant three-dimensional structure in which the atoms are covalently bonded to each other.

- Covalent network substances have high melting and boiling points because the atoms are held tightly together by strong covalent bonds and a lot of energy is needed to break the bonds.

- Ionic compounds exist as lattices in which electrostatic attractions hold the oppositely charged ions in a three-dimensional structure.

- Ionic compounds have high melting and boiling points because it takes a lot of energy to break ionic bonds.

- Solid ionic compounds do not conduct electricity because the ions cannot move.

- Ionic substances can conduct electricity when melted or in solution because the ions are free to move.

- When ionic compounds dissolve in water the electrostatic attractions between ions are replaced by attractive forces between the ions and the water molecules.

- Many ionic compounds are highly coloured compared to covalent substances.

3 Formulae and reaction quantities

You should already know

- The Group 7 elements (halogens) and hydrogen, oxygen and nitrogen are diatomic elements.
- Compounds are formed when atoms of two or more different elements join together.
- How to name a two-element compound from the names of the elements.
- The name of the compound indicates the elements present.
- Elements can be identified from a simple chemical formula.
- Chemical reactions can be described using word equations.

Learning intentions

In this chapter you will learn about:

- Chemical formulae for compounds, including those containing ions with more than one atom (group ions).
- Writing and balancing chemical equations.
- Gram formula mass and the mole.
- Calculations involving mass into moles and moles into mass.
- Connecting mass, volume of solutions, concentration and moles.
- Calculating quantities from balanced equations.
- Calculating percentage composition.

Formulae of elements and compounds

It is very important to know the rules for writing chemical formulae and symbol equations and to be able to apply the rules, in order to make progress in this part of the unit. The basic rules for formula and equation writing are summarised below with worked examples. Make sure you try the activities and check your answers before continuing with this chapter.

Formulae of elements

Most elements have their symbol as their formula, e.g. sodium: Na, copper: Cu, sulfur: S. The exceptions are the diatomic molecules: H_2, N_2, O_2, F_2, Cl_2, Br_2 and I_2.

Formulae of compounds using valency

Valency is another word for combining power. The valency of an element tells us how many bonds it can form with other atoms. The valency of a particular element can be worked out from its group number in the Periodic Table and is summarised in Table 1.3.1.

Table 1.3.1: *Working out valency*

Group	Valency	Example
1–4	same as group number	K: group 1: valency = 1 C: group 4: valency = 4
5–7	8 – group number	N: group 5: valency = 8 – 5 = 3 O: group 6: valency = 8 – 6 = 2

The noble gases have a valency of 0. Hydrogen has a valency of 1. The table below shows how valency can be used to work out the chemical formulae of simple two-element compounds.

Table 1.3.2: *Using valency*

Elements	H S	Mg Cl	Al O
Valency	1 ⟍ ⟋ 2	2 ⟍ ⟋ 1	3 ⟍ ⟋ 2
Formula ratio	2 ⟋ ⟍ 1	1 ⟋ ⟍ 2	2 ⟋ ⟍ 3
Formula	H_2S	$MgCl_2$	Al_2O_3
Name	hydrogen sulfide	magnesium chloride	aluminium oxide

If the valency is the same then the numbers cancel down. The formula for calcium oxide can be used as an example (Table 1.3.3).

Table 1.3.3: *Working out the formula for calcium oxide*

Elements	Ca O
Valency	2 ⟍ ⟋ 2
Formula ratio	$\cancel{2}$ 1 ⟋ ⟍ $\cancel{2}$ 1
Formula	CaO

The formulae of some compounds can be worked out from their names because they have a prefix to indicate the number of atoms: mono- = one; di- = two; tri- = three; tetra- = four; penta- = five, etc.

Activity 1.3.1

Work out the formulae of the following:

(a) Magnesium

(b) Chlorine

(c) Boron hydride

(d) Carbon monoxide

(e) Phosphorus trichloride

Activity 1.3.2

Write word and formulae equations for the following reactions:

(a) Hydrogen reacting with bromine to produce hydrogen bromide

(b) Phosphorus reacting with chlorine to produce phosphorus pentachloride

Make the Link

Nomenclature is often linked to linguistics and language studies, but here it can help you understand the science and make up of ionic compounds.

Examples: (1) iodine **mono**chloride = ICl

(2) silicon **di**oxide = SiO_2

Hydrogen sulfide and iodine monochloride are covalent compounds consisting of individual molecules and their formulae tell us exactly how many atoms of each element are in each molecule.

Silicon dioxide is a giant covalent network – there are no molecules. Its formula tells us the ratio of the atoms of each element in the compound – SiO_2 has a Si:O ratio of 1:2. This applies to all covalent network compounds.

Aluminium oxide is ionic and exists in a giant crystal lattice. Its formula tells us the ratio of the ions in the compound – Al_2O_3 has a Al:O ratio of 2:3. This applies to all ionic compounds.

Chemical equations

A chemical equation is a shorthand way of showing chemicals reacting and the new chemicals produced:

reactants → products

Chemical equations can be written in words and chemical formulae.

Example 1

Word equation: hydrogen + chlorine → hydrogen chloride

Formulae equation: H_2 + Cl_2 → HCl

Example 2

Word equation: hydrogen + oxygen → water

Formulae equation: H_2 + O_2 → H_2O

Chemical formulae for compounds containing ions with more than one atom

Some ions consist of more than one atom and are known as **group ions**, e.g. nitrate (NO_3^-), carbonate (CO_3^{2-}), and ammonium (NH_4^+). At this stage you do not need to know how these ions are formed but you do need to know how to work out chemical formulae using them. These group ions are given on the following page and listed in the SQA data booklet so you don't have to learn them.

Table 1.3.4: *Group ions*

One positive		One negative		Two negative		Three negative	
ion	*formula*	*ion*	*formula*	*ion*	*formula*	*ion*	*formula*
ammonium	NH_4^+	ethanoate	CH_3COO^-	carbonate	CO_3^{2-}	phosphate	PO_4^{3-}
		hydrogencarbonate	HCO_3^-	chromate	CrO_4^{2-}		
		hydrogensulfate	HSO_4^-	dichromate	$Cr_2O_7^{2-}$		
		hydrogensulfite	HSO_3^-	sulfate	SO_4^{2-}		
		hydroxide	OH^-	sulfite	SO_3^{2-}		
		nitrate	NO_3^-				
		permanganate	MnO_4^-				

Note that the name endings for all but one of the negative ions are either **–ate** or **–ite**. Hydroxide (OH^-) is the only exception. This will be the name ending in the compound they are in. Note also there is one positive ion, ammonium (NH_4^+).

The valency method can be used to work out chemical formulae for compounds containing group ions. The valency is the **number** of charges on the ion. For example, the carbonate ion (CO_3^{2-}) has a 2- charge so its valency is 2 (see Table 1.3.5).

Table 1.3.5: *Valency of sodium carbonate*

Element/ion	Na	CO_3^{2-}
Valency	1	2
Formula ratio	2	1
Formula	Na_2CO_3	

Note that the formula of the carbonate ion doesn't change – the 1 means there is one CO_3^{2-} ion in the formula.

An alternative to the valency method for working out the formulae of **ionic** compounds is balancing the charges on the ions. In the previous example, sodium forms an ion with a 1+ charge, Na^+. We know that the charge on the sodium ion is + because it is a metal ion and there is only one positive charge because the valency is 1. In sodium carbonate two sodium ions would be required to balance the 2- charge on the carbonate ion. The formula would therefore be $(Na^+)_2CO_3^{2-}$ or without ion charges, Na_2CO_3. The charges on ions, are not usually shown in the chemical formula. If you are asked to write an ionic formula only use a **bracket** if there is more than one of the ions, as in sodium carbonate.

🔍 Hint

The name ending of an ionic compound indicates which ions are in it. This is illustrated in the following examples:
- Sodium sulf**ide**: the **–ide** tells us there are two elements, one is sodium so the other must be the sulf**ide** ion – S^{2-}.
- Sodium sulf**ate**: the table shows that sulf**ate** is SO_4^{2-}.
- Sodium sulf**ite**: the table shows that sulf**ite** is SO_3^{2-}.

📖 Word bank
- **Group ions**

Ions with more than one atom, e.g. sulfate (SO_4^{2-}).

Brackets are also used when there is more than one group ion in a formula:

Example: *Magnesium nitrate*

Element/ion	Mg	NO_3^-
Valency	2	1
Formula ratio	1	2
Formula	$Mg(NO_3)_2$	

Roman numerals are used in chemical formulae to indicate the valency for elements that can have more than one valency – this is the case with many transition metals.

Example: *Copper(II) nitrate*

Element/ion	Cu	NO_3^-
Valency	2	1
Formula ratio	1	2
Formula	$Cu(NO_3)_2$	

> **🔍 Hint**
>
> It is useful to learn the Roman numerals up to 5:
> I=1; II=2; III=3; IV=4; V=5.

Chemical equations with compounds containing group ions are written in the same way as equations for simple compounds.

Example 1.3.1

Copper reacts with silver(I) nitrate solution to produce silver and copper(II) nitrate solution.

Word equation	copper + silver(I) nitrate → silver + copper(II) nitrate
Formula equation	Cu + $AgNO_3$ → Ag + $Cu(NO_3)_2$

Symbols can be used to indicate the state in which the reactants and products exist:

solid=(**s**); liquid=(**ℓ**); gas=(**g**) and solution=(**aq**).

The symbol (**aq**) comes from the Latin word '**aq**ua', which means 'water'. The equation above can be re-written to include state symbols:

$$Cu(s) + AgNO_3(aq) \rightarrow Ag(s) + Cu(NO_3)_2(aq)$$

GO! Activity 1.3.3

1. Use the chemical formulae rules to work out formulae for the following compounds. The group ions are on page 61.

    (a) sodium sulfite (b) potassium sulfate (c) calcium hydrogencarbonate (d) cobalt(III) nitrate (e) ammonium phosphate (f) magnesium hydroxide

2. Write word and formulae equations for the following reactions and include state symbols.

(a) Zinc and copper(II) sulfate solution reacting to form copper and zinc sulfate solution (zinc has a valency of 2).

(b) Barium hydroxide solution reacting with lithium sulfate solution to produce solid barium sulfate and lithium hydroxide solution.

Balancing chemical equations

If you look closely at the equations in examples 1 and 2 on page 60 you will see that the number of reactant atoms and product atoms are different.

In example 1:

H_2	+	Cl_2	→	HCl
2 hydrogen atoms		2 chlorine atoms		1 hydrogen atom 1 chlorine atom

These equations are said to be **unbalanced**. Unbalanced equations are adequate for showing what is reacting and what is produced, but they don't give a true picture of the **quantities** involved. The number of reactant and product atoms must be the same – they need to be **balanced**. The way to balance an equation is shown below.

Example 1, unbalanced equation:

$$H_2 \quad + \quad Cl_2 \quad \rightarrow \quad HCl$$

There are more hydrogen and chlorine atoms on the left-hand side of the equation than on the right-hand side. The atoms can't just 'disappear' – they have to be accounted for. The 'missing' hydrogen and chlorine atoms must have combined to form **another** hydrogen chloride molecule (as this is the only product):

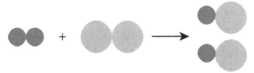

H_2 molecule	Cl_2 molecule	→	2 HCl molecules
(2 H atoms)	(2 Cl atoms)		(2H atoms + 2Cl atoms)

There are now the same number of hydrogen and chlorine atoms on each side of the equation. The equation is **balanced**. It can be written without the pictures of the atoms and molecules:

$$H_2 + Cl_2 \rightarrow 2HCl$$

Number of reactant atoms = Number of product atoms

> ## 📖 Word bank
>
> • **Balanced equation**
> The total number of atoms of each element on the left-hand side of the equation equals the total on the right.

Notice that the formulae of the reactants and products are the same as they were in the unbalanced equation. **Formulae never change**. To balance an equation numbers may only be put **in front** of formulae.

Example 2, unbalanced equation:

$$H_2 \quad + \quad O_2 \quad \rightarrow \quad H_2O$$

The molecule pictures again show the equation is unbalanced. There is one more oxygen atom on the left-hand side of the equation than on the right-hand side. It must have formed a second water molecule by reacting with another hydrogen molecule:

2 H_2 molecules   1 O_2 molecule $\rightarrow$ 2 H_2O molecules
(4 H atoms)   (2 O atoms)   (4 H atoms + 2 O atoms)

Number of reactant atoms = Number of product atoms

Without the atom pictures we have:

$$2H_2 + O_2 \rightarrow 2H_2O$$

Although in each separate equation the number of reactant and product atoms is the same, it must be remembered that they have changed the way they are arranged. Bonds between reactant molecules have been **broken** and **new bonds** have been made. In other words, **new substances** have been formed. This happens in all chemical reactions.

Equations involving compounds which have group ions in them are balanced in the same way.

Example 1.3.2	Word equation

copper + silver(I) nitrate $\rightarrow$ silver + copper(II) nitrate

Unbalanced equation

$$Cu + AgNO_3 \quad \rightarrow \quad Ag + Cu(NO_3)_2$$

The unbalanced formula equation shows that there are two NO_3 groups on the right-hand side but only one on the left-hand side. To balance the two NO_3 groups on the right-hand side a 2 has to be put in front of the $AgNO_3$ on the left-hand side.

$$Cu + 2AgNO_3 \quad \rightarrow \quad Ag + Cu(NO_3)_2$$

The equation is still not balanced because there are now two Ag on the left-hand side. This is balanced by adding a 2 in front of the Ag on the right-hand side.

Balanced equation

$$Cu + 2AgNO_3 \quad \rightarrow \quad 2Ag + Cu(NO_3)_2$$

GO! Activity 1.3.4

Balance the following equations:
1. $K + S \rightarrow K_2S$
2. $Na + Cl_2 \rightarrow NaCl$
3. $C + CO_2 \rightarrow CO$
4. $AgNO_3 + MgCl_2 \rightarrow$
   $AgCl + Mg(NO_3)_2$

Gram formula mass and the mole

The relative atomic mass of an **element** is the average mass of the isotopes of the element. The relative atomic masses of selected elements are listed in the SQA data booklet. The relative formula mass of a **compound**, often referred to as formula mass, can be calculated from the formula of a compound by adding together the relative atomic masses of all the atoms shown in the formula.

- For carbon dioxide, CO_2:
  $$C \qquad O_2$$
  The formula mass of CO_2 is **44**. $\quad 12 + (2 \times 16) = 44$

- For silver nitrate, $AgNO_3$:
  $$Ag \quad N \quad O_3$$
  The formula mass of $AgNO_3$ is **170**. $\quad 108 + 14 + (3 \times 16) = 170$

- For magnesium hydrogencarbonate, $(HCO_3)_2Mg$:
  $$(H \quad C \quad O_3)_2 \qquad Mg$$
  $$[1 + 12 + (3 \times 16)] \times 2 + 24 \cdot 5 = 146 \cdot 5$$

The formula mass of $(HCO_3)_2Mg$ is **146·5**.

In order to do chemical calculations chemists use a quantity called the **mole**, often shortened to **mol**. One mol of any substance is its formula mass in grams, i.e. the **gram formula mass**.

So, using the examples above, 1 mol of CO_2 has a gram formula mass of **44 g**; 1 mol of $AgNO_3$ has a gram formula mass of **170 g**; 1 mol of magnesium hydrogencarbonate has a gram formula mass of **146·5 g**.

> 📖 **Word bank**
> - **Gram formula mass (gfm)**
> The formula mass of a compound measured in grams.

> 📖 **Word bank**
> - **The mole**
> The gram formula mass of a substance.

Calculations involving mass into moles and moles into mass

It is possible to have more than one mole of a substance and also fractions of a mole.

For example, **2 mol** of $CO_2 = 2 \times 44 = $ **88 g** and **0·5 mol** of $CO_2 = 0 \cdot 5 \times 44 = $ **22 g**.

Generally, **mass = mol × gram formula mass**.

From this, **mol = mass/gram formula mass**.

Calculate the **mass** of $AgNO_3$ in 1·75 mol. **Example 1.3.3**

Worked answer   mass = mol × gram formula mass
$$= 1 \cdot 75 \times 170$$
$$\text{mass} = 297 \cdot 5 \text{ g}$$

Example 1.3.4

Calculate the **mass** of $(HCO_3)_2Mg$ in 0·35 mol.

Worked answer mass = mol × gram formula mass
$$= 0.35 \times 146 \cdot 5$$
$$\text{mass} = 51 \cdot 3g$$

Example 1.3.5

Calculate the number of **moles** of $AgNO_3$ in 32 g.

Worked answer mol = mass/gram formula mass
$$= 32/170$$
$$\text{mol} = 0 \cdot 19 \text{ mol}$$

Example 1.3.6

Calculate the number of **moles** of $(HCO_3)_2Mg$ in 182·9 g.

Worked answer mol = mass/gram formula mass
$$= 182 \cdot 9/146 \cdot 5$$
$$\text{mol} = 1 \cdot 25 \text{ mol}$$

GO! Activity 1.3.5

Do each question on your own but make sure you check your working and final answers with someone else in the class before going on to the next question.

1. Calculate the **gram formula mass** for each the following compounds:
   (a) magnesium fluoride (MgF_2) (b) lead(II) hydroxide $(Pb(OH)_2)$
   (c) calcium nitrate $(Ca(NO_3)_2)$ (d) iron(III) sulfate $(Fe_2(SO_4)_3)$.

Use your answers to **1.** (a)–(d) to answer questions **2.** and **3.**

2. Calculate the **mass** of each of the following:
   (a) 0·45 mol of MgF_2 (b) 1·31 mol of $Pb(OH)_2$
   (c) 0·85 mol of $Ca(NO_3)_2$ (d) 1·72 mol of $Fe_2(SO_4)_3$
3. Calculate the number of **moles** in each of the following:
   (a) 96·61 g of MgF_2 (b) 59·92 g of $Pb(OH)_2$
   (c) 234·55 g of $Ca(NO_3)_2$ (d) 92·43 g of $Fe_2(SO_4)_3$

🔍 Hint

You may wish to use the triangle to help you to remember the relationship between moles, mass and gram formula mass (gfm).

From the triangle:

mol = mass/gfm;
mass = mol × gfm.

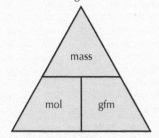

📖 Word bank

• **Standard solutions**
Solutions of accurately known concentration.

Connecting mass, volume of solutions, concentration and moles

Much of the practical work done in the laboratory and many of the reactions carried out in industry take place with the reactants dissolved in water, i.e. as aqueous solutions. These solutions have to be made up accurately and their concentration known. They are known as **standard solutions**. The following figures detail the steps which need to be followed to accurately prepare a standard solution.

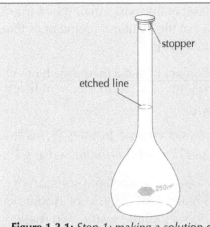

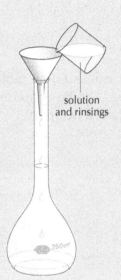

Figure 1.3.1: *Step 1: making a solution of known concentration – a volumetric flask which has a line etched (scratched) onto its long neck to indicate the exact volume is used*

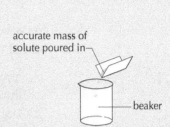

Figure 1.3.2: *Step 2: The required mass of substance to be dissolved is weighed out on a balance then dissolved in a small volume of distilled water in a beaker*

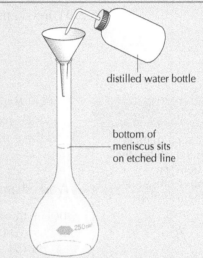

Figure 1.3.3: *Step 3: The solution is poured into the volumetric flask*

Figure 1.3.4: *Step 4: Distilled water is used to rinse all the solution out of the beaker and funnel into the flask. More water is added up to the etched line. The flask is then stoppered and inverted (turned gently upside down) several times*

Activity 1.3.6: Paired activity

You may wish to work with a partner on this activity.
Use your knowledge of chemistry and chemical techniques to answer these questions about making a standard solution.

1. Why do you think distilled water is used and not tap water?
2. What would you do to the mixture in the beaker to make sure that all the substance (solute) dissolved in the water?
3. Why do you think it is important to rinse all the solution out of the beaker?
4. Why do you think the flask is inverted several times?

Concentration (c) is a combination of the mass (m) or moles (mol) of substance and the volume (v) of the solution – remember that mass and moles are closely connected:

$$\text{concentration of solution} = \frac{\text{mass of substance dissolved}}{\text{volume of solution made (in \textbf{litres})}}$$

$$\textbf{c = m/v}$$

Mass is measured in grams (g) and volume in litres (l), so the unit of concentration is grams per litre – this is expressed as **g l⁻¹**.

Example 1.3.7

A laboratory technician was asked to make up a solution of sodium carbonate. She weighed out 10·6 g of sodium carbonate (Na_2CO_3), dissolved it in a little water then made the solution up to exactly 500 cm³.

Calculate the **concentration** of the resulting solution.

Worked answer: **c = m/v**   m = 10·6 g
                                  v = 0·5 l

So, c = 10·6/0·5
c = 21·2 g l⁻¹

Concentration is more often expressed as **mol l⁻¹**. In the previous example mass (10.6 g) can be converted to moles and the concentration then worked out from:

$$\textbf{c = mol/v}$$

Worked answer:

mol = mass/gram formula mass | gfm for Na₂ C   O₃
    = 10·6/106 | | | |
    = 0·1 mol | (2 × 23) + 12 + (3 × 16)
                                | = 106 g

Then using, c = mol/v
              = 0·1/0·5
          c = 0·2 mol l⁻¹

From **c = mol/v** we can see that **mol = c × v** and **v = mol/c**

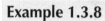
Example 1.3.8

Calculate the number of **moles** in 100 cm³ of a 0·25 mol l⁻¹ sodium hydroxide solution.

Worked answer: mol = c × v
                   = 0·25 × 0·1
               mol = 0·025 mol

Example 1.3.9

Calculate the **volume** of a 0·75 mol l⁻¹ solution which contains 0·35 mol of sodium chloride.

Worked answer: v = mol/c
                 = 0·35/0·75
             v = 0·47 l

Calculate the **concentration** of a solution, in **mol l⁻¹**, of sodium hydroxide (NaOH) made when 0·12 g is dissolved in water and made up to 200 cm³.

Example 1.3.10

Worked answer:

The answer has to be in **mol l⁻¹** so the mass in grams must first be converted to moles:

mol = m/gfm
   = 0·12/40
   = 0·003 mol

gfm for Na O H
     | | |
   23+16+1 = 40 g

Then, c = mol/v
    = 0·003/0·20
  c = 0·015 mol l⁻¹

Calculate the **mass** of copper(II) sulfate required to prepare 250 cm³ of a solution of concentration 0·15 mol l⁻¹.

Example 1.3.11

Worked answer:

In order to calculate mass in a solution the number of moles must first be calculated.

mol = c × v
   = 0·15 × 0·25
mol = 0·038 mol

Then, mass = mol × gfm
     = 0·038 × 159·5
  mass = 6·06 g

gfm of Cu S   O₄
     | | |
  63·5 + 32 + (4 × 16)
   = 159·5 g

🔘 Activity 1.3.7

Use the examples above to help you complete the following calculations:

1. Calculate the **concentration**, in **mol l⁻¹**, of the solution formed when 42·5 g of silver nitrate (gfm 170 g) is dissolved in a little water and made up to 500 cm³ with water.
2. Calculate the number of **moles** in 150 cm³ of a 0·2 mol l⁻¹ magnesium hydrogencarbonate solution.
3. Calculate the **volume** of a 0·45 mol l⁻¹ solution which contains 0·65 mol of iron(II) bromide.
4. Calculate the **mass** of ammonium nitrate (gfm 80 g) required to prepare 50 cm³ of a solution of concentration 0·9 mol l⁻¹.

Calculating quantities from balanced equations

Quantities reacting and being produced can be calculated from balanced equations. We can do this because 1 mole of any substance contains the same number of atoms in an element and molecules or ionic units in a compound. The numbers in front of a formula in an equation tell us the number of **moles** reacting and being produced. The number 1 is never written in front of a formula in an equation – the fact that a formula is written tells you there must be at least one mole present.

Using the reaction of hydrogen and chlorine to illustrate this connection:

$$\text{hydrogen} + \text{chlorine} \rightarrow \text{hydrogen chloride}$$

$$H_2 \quad + \quad Cl_2 \quad \rightarrow \quad 2HCl$$

$$1 \text{ mol} \quad 1 \text{ mol} \quad \rightarrow \quad 2 \text{ mol}$$

Converting to gram formula mass: $\quad 2 \text{ g} \quad + \quad 71 \text{ g} \quad \rightarrow \quad 73 \text{ g} \, (2 \times 35 \cdot 5)$

So, by proportion, $\quad 10 \text{ mol} \quad 10 \text{ mol} \rightarrow \quad 20 \text{ mol}$

Converting to grams: $20 \text{ g} \quad + \quad 710 \text{ g} \quad \rightarrow \quad 730 \text{ g}$

and, $\quad 0 \cdot 1 \text{ mol} \quad 0 \cdot 1 \text{ mol} \rightarrow 0 \cdot 2 \text{ mol}$

Converting to grams: $0 \cdot 2 \text{ g} + \quad 7 \cdot 1 \text{ g} \quad \rightarrow \quad 7 \cdot 3 \text{ g}$

Example 1.3.12

(a) How many moles of hydrogen chloride are produced when $0 \cdot 4$ mol of hydrogen reacts with excess chlorine? (Excess means more than enough for complete reaction.)

Worked answer: $\quad \text{hydrogen} + \text{chlorine} \rightarrow \text{hydrogen chloride}$

$$H_2 \quad + \quad Cl_2 \quad \rightarrow \quad 2HCl$$

$$1 \text{ mol} \quad 1 \text{ mol} \rightarrow \quad 2 \text{ mol}$$

Key relationship $\quad$ **1 mol** $\quad\quad\quad \rightarrow \quad$ **2 mol**

So, $\quad\quad\quad 0 \cdot 4 \text{ mol} \quad\quad\quad \rightarrow 0 \cdot 8 \text{ mol} \, (2 \times 0 \cdot 4)$

Moles of hydrogen chloride produced is **0·8 mol**.

Although not asked for in this example, the number of moles of chlorine can be worked out ($0 \cdot 4$ mol).

If the question had asked for the answer to be in grams, the number of moles is multiplied by the gram formula mass (gfm):

$\text{mass} = \text{moles} \times \text{gfm}$

$\quad\quad = 0 \cdot 8 \times 36 \cdot 5$

$\text{mass} = \textbf{28·6 g}$

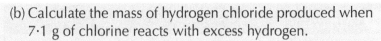

(b) Calculate the mass of hydrogen chloride produced when 7·1 g of chlorine reacts with excess hydrogen.

Worked answer: hydrogen + chlorine → hydrogen chloride

$$H_2 \quad + \quad Cl_2 \quad \rightarrow \quad 2HCl$$

1 mol   1 mol →   2 mol

Key relationship   **1 mol →   2 mol**

Converting to gram   71 g →   73 g

formula mass:

So,   7·1 g →   7·3 g

Mass of hydrogen chloride produced is **7·3 g**.

The mass of hydrogen is not asked for in this question but as in (a) it could be calculated from the balanced equation.

(c) Calculate the mass of hydrogen needed to produce 21·9 g of hydrogen chloride.

Worked answer: hydrogen + chlorine → hydrogen chloride

$$H_2 \quad + \quad Cl_2 \quad \rightarrow \quad 2HCl$$

1 mol   1 mol →   2 mol

Key relationship   **1 mol**   →   **2 mol**

Converting to gram   2 g   →   73 g

formula mass:

So, 2/73 g   ←   1 g (73/73)

So, 2/73 × 21·9 = 0·6 g ←   21·9 g (21·9 × 1)

Mass of hydrogen needed is **0·6 g**.

In this calculation the arrows point right to left in the part of the calculation showing the mass of hydrogen needed is calculated backwards from the mass of hydrogen chloride produced.

Example 1.3.13

Calculate the mass of silver metal produced when 3·2 g of Cu is added to excess silver(I) nitrate solution.

Worked answer: Cu   + $2AgNO_3 \rightarrow 2Ag + Cu(NO_3)_2$

          1 mol   2 mol   2 mol   1 mol

Key relationship **1 mol**   **→ 2 mol**

Converting to gram
formula mass: 63·5 g   → 216 g (2 × 108)

     so,   1 g (63·5/63·5)   → 3·4 g (216/63·5)

     so,   3·2 g (1 × 3·2)   → 10·9 g (3·4 × 3·2)

The mass of silver produced is **10·9 g**.

> **GO! Activity 1.3.8**
>
> Use the previous examples to help you complete the following calculations:
>
> 1. Calculate the mass of water produced when 8 g of hydrogen burns in oxygen.
>
>   Balanced equation: $2H_2(g) + O_2(g) \rightarrow 2H_2O(\ell)$
>
> 2. Calculate the mass of iron(III) sulfide produced when 2.8 g of iron reacts with excess sulfur.
>
>   Balanced equation: $2Fe + 3S \rightarrow Fe_2S_3$

Calculating percentage composition

The percentage composition of an element in a compound can be calculated using the following relationship:

$$\text{Percentage composition by mass} = \frac{\text{mass of the element in formula}}{\text{formula mass of compound}} \times 100$$

Example 1.3.14

Calculate the percentage composition of iron in iron(III) sulphide (Fe_2S_3).

Worked answer: Percentage composition of iron

$$= \frac{\text{mass of iron in formula}}{\text{formula mass of iron (III) sulphide}} \times 100$$

$$= \frac{(3 \times 56)}{(3 \times 56) + (2 \times 32)} \times 100$$

$$= \frac{168}{232} \times 100$$

Percentage composition of iron = 72.4%.

More examples can be found in Area 3. Make sure you try Activities 3.7.1 and 3.9.4.

Learning checklist

In this chapter you have learned:

- How to write chemical formulae for compounds, including those containing ions with more than one atom (group ions).
- How to write formulae equations for compounds, including those containing ions with more than one atom (group ions).
- How to balance chemical equations.
- The connection between gram formula mass and the mole.
- How to carry out calculations involving mass into moles and moles into mass.

  mass = mol × gram formula mass

  mol = mass/gram formula mass

- The connection between mass, volume of solutions, concentration and moles.

  c = mol/v; mol = c × v; v = mol/c

- How to calculate quantities from balanced equations.
- How to calculate percentage composition.

4 Acids and bases

Make the Link

Vitamin C (ascorbic acid) is one of several essential vitamins and nutrients required by the human body. The study of human nutrition is covered in Health and Food Technology courses.

Acids and bases all around

Acids and bases are all around us in our everyday lives. They are present in the foods we eat, the medicines we take and are essential for our body to stay healthy and work properly. Acids are used as additives to drinks and foods as they alter their taste and act as preservatives. Phosphoric acid and carbonic acid are present in many cola drinks. Citric acid and ethanoic acid (vinegar) are used for flavouring and preserving food. Ascorbic acid (vitamin C) is an essential vitamin required by our body and is found in citrus fruits. Acetylsalicylic acid (aspirin) is used as a pain killer and for reducing fever. Hydrochloric acid is present in our stomach and helps break down food. In industry nitric acid reacts with ammonia, a **base**, to make ammonium nitrate fertiliser (see Chapter 9: Fertilisers). Sweet smelling esters used in the perfume industry are made when carboxylic acids react with alcohols.

Magnesium hydroxide and calcium carbonate are bases found in indigestion remedies. They neutralise excess acid in the stomach. Bicarbonate of soda (sodium hydrogencarbonate) is a base used in baking as a raising agent. It reacts with acids like lemon juice to produce carbon dioxide gas which bubbles through the baking product helping it rise. Ammonia is a good solvent for oil and grease and is used in some kitchen cleaning products, especially when a streak-free finish is desired. Some oven cleaners and paint strippers contain concentrated bases. Sodium hydroxide is used to make soap and in the papermaking industry.

STEP BACK IN TIME: ACIDS AND ALKALIS

Acids and alkalis (soluble bases) have been known since ancient times. The word acid comes from the Latin word acere, *which means 'sour'. All acids taste sour. Acid tastes which have been known for thousands of years are vinegar (see Chapter 6: Everyday consumer products: Carboxylic acids), sour milk and lemon juice. The word alkaline comes from the Arabic al-qily, which means 'to roast in a pan' and relates to alkaline substances found in the ashes of plants that have been burned. Mixing the ashes with water produces a mixture of potassium and sodium carbonates in solution.*

The making of acids was recorded as far back as the eighth century in the Middle East. Sulfuric acid was made by heating green vitriol *(iron(II) sulfate) and condensing the vapour into water. Mixing a vitriol with* nitre *(potassium nitrate) and heating produced vapours which gave nitric acid. Adding* sal ammoniac *(ammonium chloride) to nitric acid gave* aqua regia *(royal water), so named for its ability to dissolve gold. Hydrochloric acid ('spirit of salt' – a name still used in commerce/pharmacy as late as the early 1970s) was also known in the Middle Ages.*

In the eighteenth century the French scientist Antoine Lavoisier proposed, wrongly, that all acids contained oxygen. In 1772 the English scientist Joseph Priestly made hydrogen chloride gas, which dissolved in water to form hydrochloric acid – he called it muriatic acid *from the Latin word* muria, *which means brine (salt water). It wasn't until the start of the nineteenth century that another English scientist, Sir Humphry Davy, proved a connection between compounds containing hydrogen and acidity.*

Word bank

- **Base**

Bases are metal oxides, hydroxides, carbonates and ammonia. Soluble bases dissolve in water to form alkaline solutions.

Figure 1.4.1: *Common foodstuffs containing acids*

Figure 1.4.2: *Bases are found in indigestion remedies and household cleaning products*

GO! Activity 1.4.1

Summarise the information about common acids and alkalis in a table. Choose suitable headings to describe the chemical and its uses.

Figure 1.4.3: *Jabir ibn Hayyan (Geber), considered by some to be the 'father of chemistry', introduced a scientific and experimental approach to early chemistry*

Figure 1.4.4: *The English chemist Sir Humphry Davy, who linked acidity with the presence of hydrogen in acidic compounds, working in his laboratory*

At the end of the nineteenth century the Swedish scientist Svante Arrhenius defined an acid as a substance which, when dissolved in water, increases the concentration of hydrogen ions. He defined bases as substances which, when added to water, increase the concentration of hydroxide ions. In 1923 another Swedish chemist, Johannes Brønsted, and the English chemist Thomas Lowry defined acids and bases depending on whether a substance gave away (donated) a proton (hydrogen ion) or accepted a proton. According to the Brønsted-Lowry theory, an acid is a proton donor and a base is a proton acceptor. These definitions are still used today.

The dissociation of water molecules

Water is described as a covalent molecular liquid (see Area 1, Chapter 2: Atomic structure and bonding: Structure and properties of covalent substances) and so should not conduct electricity. However, if a very sensitive conductivity meter is used a small current is detected when electricity is passed through pure water. Pure water is tested as opposed to tap water, as tap water has natural salts dissolved in it as well as chemicals added at the water treatment plant. This result suggests that there must be ions present in pure water. The ions result from the dissociation (breaking down) of some of the water molecules into equal numbers of hydrogen and hydroxide ions. This dissociation can be represented as:

$$H_2O(\ell) \rightleftharpoons H^+(aq) + OH^-(aq)$$

The $\rightleftharpoons$ symbol indicates that the reaction is reversible and occurs in both directions. As water molecules dissociate into ions at the same time most of the ions recombine to form molecules. It has been calculated that for every hydrogen and hydroxide ion there are 555 million water molecules. The (aq) symbol indicates that the ions are surrounded by water molecules. The pH of pure water is 7, which indicates it is neutral – neither acid nor alkali.

Word bank

• **Dissociation of water**

A small proportion of water molecules break down into equal concentrations of hydrogen ions and hydroxide ions.

pH and hydrogen ion concentration

Every solution made by dissolving a substance in water (aqueous solution) contains hydrogen ions, $H^+(aq)$, and hydroxide ions, $OH^-(aq)$. Some substances dissolve in water and change the concentrations of hydrogen and hydroxide ions in solution. When the hydrogen ion concentration of a solution increases the hydroxide ion concentration decreases and vice versa. The pH scale is a convenient way of indicating the hydrogen ion concentration of a solution.

Pure water has an equal concentration of hydrogen ions and hydroxide ions and has a pH of 7 and is described as neutral.

An acid is formed when a substance dissolves in water, increasing the concentration of hydrogen ions, $H^+(aq)$, in the solution. Acid solutions have a higher concentration of $H^+(aq)$ than water and have pH values of less than 7. The greater the hydrogen ion concentration, the lower the pH. An acid with a pH of 1 has a high hydrogen ion concentration. An acid with a pH of 6 would have a lower hydrogen ion concentration than an acid with pH 1. Both acids will still contain contain $OH^-(aq)$ but the concentration will be less than in pure water.

An alkali is formed when a substance dissolves in water increasing the concentration of hydroxide ions, $OH^-(aq)$ in the solution. Again, there will be $H^+(aq)$ present but the concentration will be lower than in pure water. The pH scale is still used – remember pH is a measure of hydrogen ion concentration. Alkaline solutions have a lower concentration of hydrogen ions than water and have a pH greater than 7. As the hydroxide ion concentration increases the hydrogen ion concentration decreases. The larger the pH value the lower the concentration of hydrogen ions and the higher the concentration of hydroxide ions. A solution of pH 14 will have a much higher concentration of hydroxide ions than a solution of pH 8.

Figure 1.4.5 summarises the comparison of ion concentration in acid and alkaline solutions with water. Note that the pH scale can have values less than 1 and greater than 14.

🔍 Hint

The symbol often used to represent concentration is [ ].

In water and neutral solutions:

$[H^+(aq)] = [OH^-(aq)]$

In acid:

$[H^+(aq)] > [OH^-(aq)]$

In alkali:

$[H^+(aq)] < [OH^-(aq)]$

NOTE:

> means 'greater than'

< means 'less than'

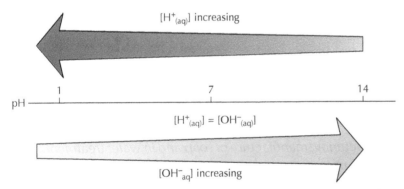

Figure 1.4.5: *Comparing $H^+(aq)$ and $OH^-(aq)$ ion concentrations in acid and alkaline solutions with water ([ ] = concentration)*

Soluble non-metal oxides dissolve in water to form acid solutions. This means that they must increase the hydrogen ion concentration in water.

Carbon dioxide in the atmosphere dissolves in rainwater making it acidic. 'Fizzy' (carbonated) drinks contain dissolved carbon dioxide which contributes to the acidity of the drink. Carbon dioxide forms carbonic acid. Some of the carbon dioxide

Make the Link

Acid rain and its impact on the Earth and the environment is covered in geography.

molecules dissolve in water to give hydrogen ions and carbonate ions. This increases the concentration of hydrogen ions in solution:

$$\text{carbon dioxide} + \text{water} \rightarrow \text{carbonic acid}$$
$$CO_2(g) + H_2O(\ell) \rightarrow 2H^+(aq) + CO_3^{2-}(aq)$$

Some common acids such as hydrochloric acid are completely ionised in solution. Hydrochloric acid is formed when the covalent gas hydrogen chloride dissolves in water. The covalent bond between the hydrogen and chlorine is broken and hydrogen and chloride ions are formed. All the hydrogen chloride molecules break down to form ions. Water molecules are attracted to each ion. The process is summarised in the equation:

$$HCl(g) + H_2O(\ell) \rightarrow H^+(aq) + Cl^-(aq)$$

Soluble metal oxides, like sodium oxide, ionise completely in water and form alkaline solutions. This must mean they increase the concentration of hydroxide ions in the water. The oxide ion from the metal oxide accepts a hydrogen from a water molecule to form a hydroxide ion:

$$O^{2-}(s) + H_2O(\ell) \rightarrow OH^-(aq) + OH^-(aq)$$

The equation has been written with two separate hydroxide ions to emphasise that the oxide ion becomes part of a hydroxide ion when it accepts a hydrogen from a water molecule.

SPOTLIGHT ON THE CHLOR-ALKALI INDUSTRY

The main products of the chlor-alkali industry are chlorine and sodium hydroxide.

Sodium hydroxide is the most important base produced and used in the chemical industry. Over 60 million tonnes are produced worldwide every year. It is used in cleaning products, in cotton textile processing, pulp and paper manufacturing, manufacture of soap and in water treatment. It is also used as a catalyst in the manufacture of biodiesel from vegetable oil.

Sodium hydroxide is used in the home as a type of drain cleaner to unblock clogged drains, usually in the form of a dry crystal or as a thick liquid gel. The alkali dissolves greases to produce water soluble products. It also breaks down the proteins, such as those found in hair which may block water pipes. These reactions are speeded up by the heat generated when sodium hydroxide and the other chemicals in the cleaner dissolve in water. These alkaline drain cleaners are highly corrosive and should be handled with great care.

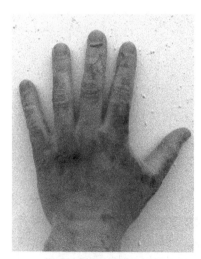

Figure 1.4.6: *Sodium hydroxide can cause severe burns and should be handled with care*

Sodium hydroxide is used in some 'relaxers' to straighten hair. However, because of the high incidence and intensity of chemical burns, manufacturers of chemical relaxers use other alkaline chemicals in preparations available to the public. Sodium hydroxide relaxers are still available, but they are used mostly by professionals.

A solution of sodium hydroxide in water was traditionally used as the most common paint stripper on wooden objects. Its use has become less common, because it can damage the wood surface, raising the grain and staining the colour.

The main way of producing sodium hydroxide is by passing electricity through (electrolysing) a concentrated sodium chloride (brine) solution. Sodium chloride is a relatively cheap and readily available resource although the huge amount of electricity used in the process is very costly. There are three types of electrolytic cell: membrane and diaphragm, which are very similar to each other, and mercury.

A diagram of a diaphragm cell is shown in Figure 1.4.7.

Chlorine gas is produced at the positive electrode and hydrogen at the negative. The two gases can react explosively together so have to be kept apart. The removal of hydrogen ions from the solution increases the concentration of the hydroxide ions in the cathode compartment. The cathode compartment eventually has about 10% sodium hydroxide and 15% sodium chloride in solution. As the water is evaporated off most of the sodium chloride crystallises out and is removed, leaving a concentrated solution of sodium hydroxide behind. Further evaporation of water leaves solid sodium hydroxide, with a small amount of sodium chloride impurity.

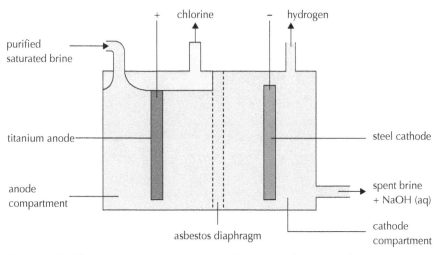

Figure 1.4.7: *The electrolysis of concentrated brine to make sodium hydroxide (anode = + electrode; cathode = − electrode)*

The diaphragm is made of porous asbestos and keeps the two compartments separate. It allows the brine to seep into the cathode compartment but prevents the hydroxide from getting into the anode compartment. The asbestos diaphragm has to be replaced every so often, which is causing some environmental issues. Asbestos is a danger to health and has to be handled and disposed of very carefully. Safe disposal of asbestos is very expensive. Membranes are replacing asbestos diaphragms, not only because of the issues of disposal of the diaphragm but also purer sodium hydroxide is produced.

At one time most of the sodium hydroxide in Britain was produced using a cell with a mercury cathode. This results in sodium forming and dissolving in the mercury. The mercury is continuously removed and passed through water where the sodium reacts to form a very pure solution of sodium hydroxide. Mercury however is very toxic and cannot be allowed to escape into the environment. In the 1950s mercury used as a catalyst in a chemical process escaped into the sea at Minamata Bay in Japan and got into the human food chain. As a result, over the years thousands of people died or suffered brain damage due to mercury poisoning. This led to a sharp decrease in using mercury and in Britain there are no large factories now producing chlorine and sodium hydroxide using a mercury cell.

The chlorine produced is as important a chemical as sodium hydroxide. It is used to make bleach, hydrochloric acid, plastics (PVC (polyvinylchloride)) and a range of solvents. The hydrogen produced is not wasted and is often used as a fuel to supply energy for various parts of the process.

 Hint

You can find out more about Minamata disease by entering the name into an internet search engine.

 Activity 1.4.3

1. Sodium hydroxide can be used to 'scrub' acidic gases and stops them escaping into the atmosphere. Carbon dioxide reacts with sodium hydroxide to form sodium carbonate and water. Write a balanced equation for the reaction.

2. Give one major advantage and disadvantage the mercury cell had over the diaphragm cell.

3. Summarise the cost implications of producing sodium hydroxide by the electrolysis of brine (sodium chloride).

4. The flow diagram shows the production of sodium hydroxide from brine. Name products A and B and processes X and Y.

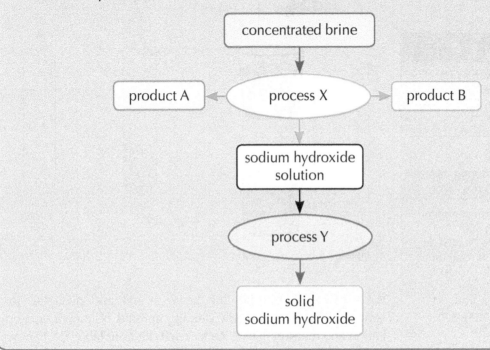

Diluting acids and alkalis

When water is added to an acid, although the number of ions in solution stays the same, the volume of the solution increases and the ions spread out through the solution. The acid has been diluted – its concentration has decreased. This means the concentration of the $H^+(aq)$ ions has decreased. It is like diluting orange squash. If a glass is quarter filled with orange then filled up with water we have diluted orange squash. The glass contains as much orange squash as before but it is now spread through a larger volume. The concentration of the orange has decreased (see figure 1.4.8).

Table 1.4.1

[H⁺(aq)]/mol l⁻¹	pH
0·1	1
0·01	2
0·001	3
0·0001	4
0·00001	(a)
0·000001	(b)

GO! Activity 1.4.4

Predict the pH values for (a) and (b) by looking at the pattern of values in Table 1.4.1.

🔍 Hint

Remember that concentration is a combination of the quantity of solute dissolved and the volume of solution formed: c = mol (or mass)/ volume (see Chapter 3: Formulae and reaction quantities: Connecting mass, volume of solutions, concentration and moles). If the volume is increased and the mass kept the same then the concentration decreases.

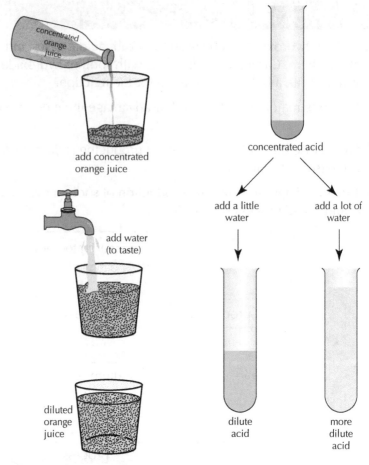

Figure 1.4.8: *Diluting acid*

We can investigate the link between pH and hydrogen ion concentration by repeatedly diluting an acid of known concentration and measuring its pH after each dilution. Table 1.4.1 shows the pH values obtained when hydrochloric acid with a concentration of $0·1$ mol l⁻¹ ($0·1$ mol l⁻¹ HCl(aq)) is repeatedly diluted by a factor of 10. $0·1$ mol l⁻¹ HCl(aq) contains $0·1$ mol l⁻¹ H⁺(aq) and $0·1$ mol l⁻¹ Cl⁻ (aq).)

Repeated dilution of a $0·1$ mol l⁻¹ sodium hydroxide solution ($0·1$ mol l⁻¹ NaOH⁻(aq)) would follow a similar pattern. The solutions would become less and less alkaline as the concentration of OH⁻(aq) ions decreased. Remember, although an alkali has a higher concentration of OH⁻(aq) ions than pure water and acids there are still H⁺(aq) ions present so pH can still be measured.

Figure 1.4.9 summarises the relationship between pH and hydrogen ion concentration and gives the pH of some everyday solutions. The hydrogen ion concentration here is given relative to distilled ('pure' water) to make comparison easier. Water is given the value 1 and the numbers given for the hydrogen ion concentration are compared to this. For example, baking soda (pH 8) has one-tenth the concentration of hydrogen ions as water whereas milk (pH 6) has ten times the concentration of hydrogen ions as water.

Concentration of hydrogen ions compared to distilled water	pH	Common solutions
1/10 000 000	14	Liquid drain cleaner, caustic soda
1/1 000 000	13	Bleaches, oven cleaner
1/100 000	12	Soapy water
1/10 000	11	Household ammonia
1/1 000	10	Milk of magnesium
1/100	9	Toothpaste
1/10	8	Baking soda, seawater, eggs
1	7	'Pure' water
10	6	Urine, milk
100	5	Acid rain, black coffee
1 000	4	Tomato juice
10 000		Grapefruit and orange juice, soft drinks
100 000	2	Lemon juice, vinegar
1 000 000	1	Hydrochloric acid secreted from the stomach lining
10 000 000	0	Battery acid

Figure 1.4.9: *pH of some common solutions and their hydrogen ion concentration*

What reacts in a neutralisation reaction?

When an acid is neutralised the hydrogen ions are removed and replaced in solution by metal ions. Acids can be neutralised using bases which include metal oxides, metal hydroxides and metal carbonates. When the base used to neutralise the acid is a metal oxide or a metal hydroxide, water and a salt are formed in the reaction. When the base used is a metal carbonate water and a salt are again formed and carbon dioxide gas is also given off. Looking at the equations helps us see exactly what is happening during the reactions. Note that in the ionic equations the ions in solution are shown separately.

Acid + metal hydroxide

sulfuric acid	+	sodium hydroxide	→	water	+	sodium sulfate
$H_2SO_4(aq)$	+	$2NaOH(aq)$	→	$H_2O(\ell)$	+	$Na_2SO_4(aq)$

Ionic equation

$$2H^+(aq) + SO_4^{2-}(aq) + 2Na^+(aq) + 2OH^-(aq) \rightarrow 2H_2O(\ell) + 2Na^+(aq) + SO_4^{2-}(aq)$$

📖 Word bank

• **Spectator ions**

Ions that do not take part in a reaction.

Looking closely at both sides of the equation it can be seen that SO_4^{2-} and $2Na^+$ appear on both sides of the equation. This means that they have not taken part in the reaction. Ions which do not take part in a reaction are known as **spectator ions**. The ions which have taken part in the reaction are the $H^+(aq)$ and the OH^- ions. Water is the only new product. Rewriting the equation without the spectator ions:

$$H^+(aq) + OH^-(aq) \rightarrow H_2O(\ell)$$

The equation clearly shows the $H^+(aq)$ ions being removed from solution as $H_2O(\ell)$.

hydrochloric acid	+	potassium hydroxide	→	water	+	potassium chloride
$HCl(aq)$	+	$KOH(aq)$	→	$H_2O(\ell)$	+	$KCl(aq)$

Ionic equation

$$H^+(aq) + Cl^-(aq) + K^+(aq) + OH^-(aq) \rightarrow H_2O(\ell) + K^+(aq) + Cl^-(aq)$$

The ionic equation shows that $Cl^-(aq)$ and $K^+(aq)$ are spectator ions. The ions which have taken part in the reaction are H^+ and OH^-. Water is the only new product. Rewriting the equation without the spectator ions:

$$H^+(aq) + OH^-(aq) \rightarrow H_2O(\ell)$$

Again, the equation clearly shows the $H^+(aq)$ being removed from solution as $H_2O(\ell)$.

Acid + metal carbonate

nitric acid   +   lithium carbonate   →   water   +   lithium nitrate   +   carbon dioxide

$2HNO_3(aq)$ + $Li_2CO_3(aq)$ → $H_2O(\ell)$ + $2LiNO_3(aq)$ + $CO_2(g)$

Ionic equation

$$2H^+(aq) + 2NO_3^-(aq) + 2Li^+(aq) + CO_3^{2-}(aq) \rightarrow H_2O(\ell) + 2Li^+(aq) + 2NO_3^-(aq) + CO_2(g)$$

The ionic equation shows that $NO_3^-(aq)$ and $Li^+(aq)$ are spectator ions. The ions which have taken part in the reaction are the $2H^+$ and the $CO_3^{2-}(aq)$ ions. Water and carbon dioxide are the only new products. Rewriting the equation without the spectator ions:

$$2H^+(aq) + CO_3^{2-}(aq) \rightarrow H_2O(\ell) + CO_2(g)$$

Again, the equation clearly shows $H^+(aq)$ being removed from solution as $H_2O(\ell)$.

Each of the acid reactions with a base clearly shows that in a neutralisation the only reaction taking place is the removal of the hydrogen ions as water.

Volumetric titrations

A **neutralisation** reaction between an acid and an alkali can be carried out accurately in an experiment called a **titration**. A titration involves a burette which is used to gradually add accurately measured volumes of the acid into a conical flask containing alkali. The alkali is accurately measured using a pipette. A chemical called an indicator is also added to the flask. Indicators change colour just as the neutralisation reaction is complete, i.e. at the end-point of the reaction. This is the signal to stop the titration. The volume of acid added to neutralise the alkali is then read off the scale on the side of the burette. The volumes of acid and alkali can then be used to calculate the unknown concentration of an acid or alkali. The details of how to carry out a titration are covered in Chapter 11: Chemical analysis.

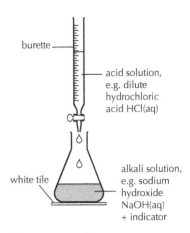

burette

acid solution, e.g. dilute hydrochloric acid HCl(aq)

white tile

alkali solution, e.g. sodium hydroxide NaOH(aq) + indicator

Figure 1.4.10: *Carrying out an acid–alkali titration*

📖 Word bank

- **Neutralisation**

When an acid reacts with a base to form water.

Calculating concentration

The concentration of an acid or alkali can be calculated using the results of a titration.

Example 1.4.1

16.0 cm³ of 0.10 mol⁻¹ sulfuric acid was required to neutralise 20.0 cm³ of potassium hydroxide solution. Find the concentration of the potassium hydroxide solution.

The balanced equation for the reaction is:

$$2KOH(aq) + H_2SO_4(aq) \rightarrow K_2SO_4(aq) + 2H_2O(\ell)$$

Worked answer: The equation tells us that 1 mol of sulfuric acid will neutralise 2 mol of sodium hydroxide. These are known as balancing numbers.

We can use this and the relationship:

$$\frac{(\text{volume} \times \text{concentration})_{alkali}}{\text{balancing no. }_{alkali}} = \frac{(\text{volume} \times \text{concentration})_{acid}}{\text{balancing no. }_{acid}}$$

$$\frac{20 \times C}{2} = \frac{16 \times 0.1}{1}$$

(Note: you don't have to change volumes to litres.)

$$10 \times C = 1.6$$

$$\mathbf{C = 0.16 \ mol \ l^{-1}}$$

Calculating volume

Example 1.4.2

0.20 mol l^{-1} of nitric acid was neutralised by 15.8 cm³ of 0.25 mol l^{-1} sodium hydroxide.

The balanced equation is:

$$HNO_3(aq) + NaOH(aq) \rightarrow H_2O(\ell) + NaNO_3(aq)$$

Calculate the volume of nitric acid required.

Worked answer: The equation tells us that both the acid and alkali have a balancing number of 1.

We can use this and the relationship:

$$\frac{(\text{volume} \times \text{concentration})_{alkali}}{\text{balancing no. }_{alkali}} = \frac{(\text{volume} \times \text{concentration})_{acid}}{\text{balancing no. }_{acid}}$$

$$\text{So, } \frac{V \times 0.2}{1} = \frac{15.8 \times 0.25}{1}$$

$$V \times 0.2 = 3.95$$

$$\text{So, } V = \frac{3.95}{0.2}$$

$$\mathbf{V = 19.8 \ cm^3}$$

Activity 1.4.5

1. The concentration of ethanoic acid in vinegar can be found by neutralising the vinegar using sodium hydroxide solution.

$$CH_3COOH(aq) + NaOH(aq) \rightarrow CH_3COONa(aq) + H_2O(\ell)$$

A sample of vinegar was neutralised using 0·80 mol l^{-1} sodium hydroxide solution. 23·6 cm^3 of sodium hydroxide solution was required to neutralise 25·0 cm^3 of vinegar.

Calculate the concentration of ethanoic acid in the vinegar.

2. 0.1 mol l^{-1} of sulfuric acid was netutralised by 17.8 cm^3 of 0.15 sodium hydroxide solution. The balanced equation is:

$$H_2SO_4(aq) + 2NaOH(aq) \rightarrow Na_2SO_4(aq) + H_2O(\ell)$$

Calculate the volume of sulfuric acid required.

Preparing soluble salts

Acid + alkali

At the point where an acid is neutralised by an alkali, the salt and water formed during the reaction and the water from the reacting solutions are present along with the indicator. If the volumes of acid and alkali which have reacted at this point are noted, the reaction could be repeated without the indicator and a pure solution of the salt in water would be obtained. The solution could then be heated and the water allowed to evaporate off leaving a pure, dry sample of the salt behind.

Acid + insoluble base

It is often more convenient to react an insoluble base with an acid to make a desired salt. This is because the base can be added to the acid in excess. The acid will react with the base until all the acid is reacted, i.e. the acid is neutralised. All of the $H^+(aq)$ ions from the acid have been taken out of solution as water. All that is left in the mixture is the salt, water and excess base. The excess base can be filtered out and the remaining solution heated to boil off most of the water then left to allow the rest of the water to evaporate and leave the pure salt behind. There is no need to measure out accurate volumes of reactants or use an indicator.

Copper chloride can be made by reacting excess insoluble copper(II) carbonate with hydrochloric acid, as shown in Figure 1.4.11.

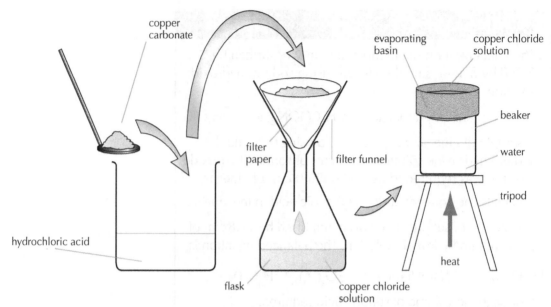

Figure 1.4.11: *Making copper(II) chloride using insoluble copper(II) carbonate*

⊙ Activity 1.4.6

The word equation for the formation of copper(II) chloride using copper(II) carbonate is shown below:

copper(II) carbonate + hydrochloric acid → copper(II) chloride + water + X

(a) Name substance X.
(b) Write a balanced equation for the reaction.
(c) An indication that the acid has been neutralised is that unreacted copper(II) carbonate would be left in the beaker.
     What else would be observed indicating that the reaction had stopped?

Learning checklist

In this chapter you have learned:

- Pure water contains H_2O molecules and equal concentrations of $H^+(aq)$ and $OH^-(aq)$ ions.

- All aqueous solutions contain $H^+(aq)$ and $OH^-(aq)$ ions.

- pH is a measure of the $H^+(aq)$ ion concentration in a solution.

- The concentration of $H^+(aq)$ ions in acids is higher than in water.

- The concentration of $H^+(aq)$ ions in alkalis is lower than in water.

- The concentration of $OH^-(aq)$ ions in alkalis is higher than in water.

- The concentration of $OH^-(aq)$ ions in acids is lower than in water.

- Diluting acids reduces the concentration of $H^+(aq)$ so the pH of the acid increases towards 7.

- Diluting alkalis reduces the concentration of $OH^-(aq)$ so the pH of the alkali decreases towards 7.

- Only the $H^+(aq)$ ions and $OH^-(aq)$ ions react in a neutralisation reaction.

- A neutralisation reaction can be followed by volumetric titration.

- Indicators are used in an acid–alkali titration to show when neutralisation has occurred.

- The results from titration can be used to calculate unknown concentrations and volumes.

- Soluble salts can be made by titrating acids and alkalis or reacting insoluble bases with acids.

5. **Homologous series**
   - Alkanes
   - Cycloalkanes
   - Alkenes

6. **Everyday consumer products**
   - Alcohols
   - Energy from fuels
   - Carboxylic acids

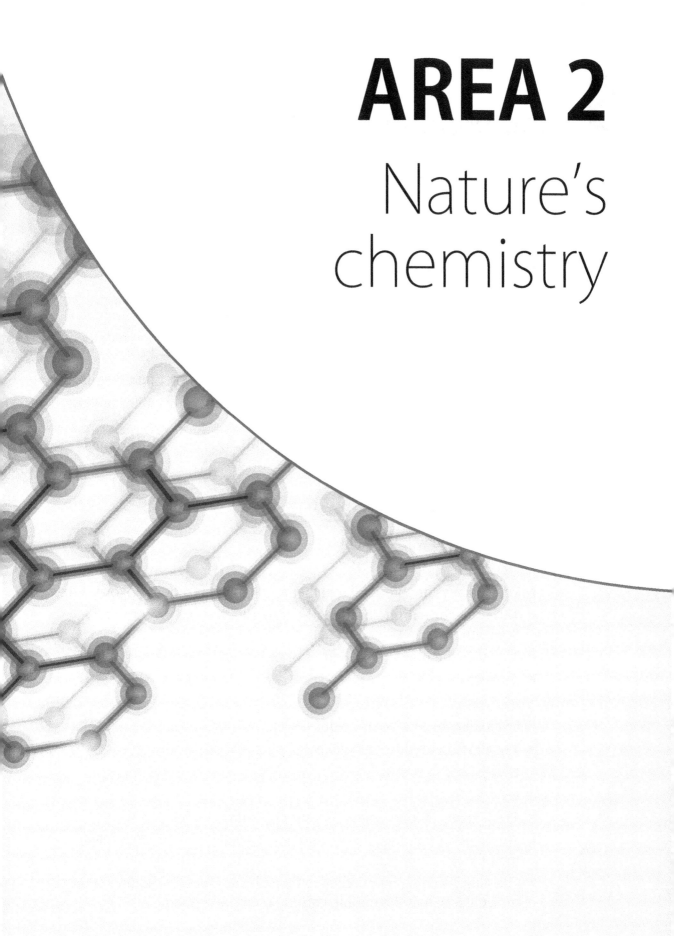

AREA 2
Nature's chemistry

5 Homologous series

SPOTLIGHT ON FUELLING SCOTLAND'S FUTURE

As North Sea gas and oil runs out Scotland will become dependent on imported gas unless new sources of gas can be identified and exploited.

In the second half of the nineteenth century and first half of the twentieth century Scotland had a thriving shale oil industry. The world's first oil refinery had been set up in 1851 at Bathgate in West Lothian. Oil was extracted from oil-bearing shale by crushing the shale rock and heating it in a retort using a process developed by James 'Paraffin' Young, a Glasgow chemist. It was not until 1859 that Americans struck free-flowing oil. Oil shale was mined in the Lothians until 1962 but the industry died out because it could no longer compete with imported oil.

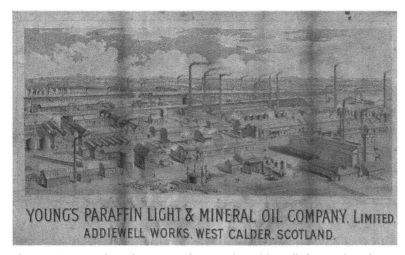

Figure 2.5.1: *Another of Young's refineries, the Addiewell chemical works in West Lothian, circa 1880 © Almond Valley Heritage Trust*

Now attention is returning to the area where the oil was first discovered. More than 20 000 square kilometres covering the entire central belt and a part of the south-west of Scotland, have been earmarked by the UK Government for possible exploitation by technologies such as fracking to extract gas from wells dug deep into the ground.

Fracking, short for hydraulic fracturing, involves drilling down to the rocks and then using explosives (or injecting high pressure water) to fracture the rocks and extract gas and oil from the shale beds. Methane can also be released from coal seams by fracking. Over 60% of gas used in the United States is obtained in this way. In 1996 America was extracting 0·3 trillion cubic feet of shale gas. By 2011 this had risen to 7·8 trillion cubic feet.

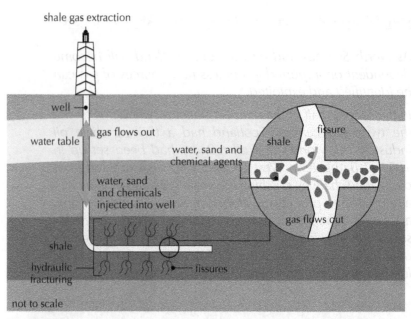

Figure 2.5.2: *Shale gas can be extracted by hydraulic fracturing, 'fracking', of shale rock*

It is thought that globally fracking could extend the world's recoverable gas reserves from 120 years to 250 years. It is estimated that reserves in China could fuel the country for the next 200 years.

The developments are controversial. In 2010 pilot fracking projects in Lancashire, England, were blamed for causing small earthquakes. This led to the projects being halted. There is also concern that chemicals used in fracking are potentially hazardous to health. To fracture the rock as much as 20 000 cubic metres of water, 1800 tonnes of sand and 100 tonnes of other chemicals need to be injected into each borehole. When the rock has been fractured the waste water is pumped out of the borehole. There is concern that the water supply could become contaminated with many of the chemicals that are used in the process. Some of these chemicals are thought to disrupt hormone levels in the body and could lead to more girls being born than would normally be the case. Some are also carcinogenic, increasing cancer risks.

Fracking might also hinder the development of green renewable technologies and lead to failure to meet climate change targets.

The photograph opposite, taken from space, shows what looks like the bright lights of a city in the middle of North Dakota. In reality, the light is caused by flaring off unwanted gas that is being produced with oil by fracking the Bakken shale beds. The low price of natural gas makes it uneconomical to build storage facilities and pipelines to transport the gas. Disposing of the gas by flaring sends more unwanted carbon dioxide into the atmosphere.

Hint

An internet search using the term 'fracking' will allow you to find out up-to-date information.

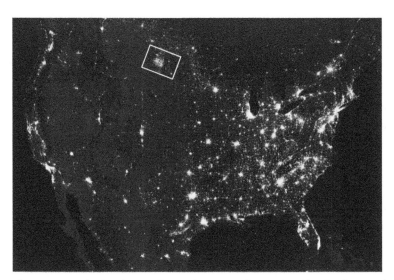

Figure 2.5.3: *The bright smudge in the rectangle is caused by the flaring of unwanted gas from the fracking process*

Alkanes

The fuels obtained by the fractional distillation of crude oil (see Fig. 2.5.4) are mainly made up of alkanes.

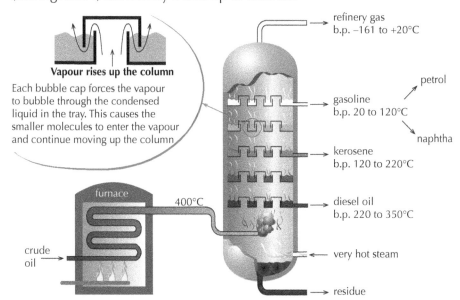

Vapour rises up the column

Each bubble cap forces the vapour to bubble through the condensed liquid in the tray. This causes the smaller molecules to enter the vapour and continue moving up the column

refinery gas
b.p. −161 to +20°C

petrol

gasoline
b.p. 20 to 120°C

naphtha

kerosene
b.p. 120 to 220°C

diesel oil
b.p. 220 to 350°C

furnace

400°C

crude oil

very hot steam

residue

Figure 2.5.4: *Crude oil is separated into fractions in a fractionating column. There are trays with bubble caps at different levels. Different fractions condense out at different levels*

Natural gas is mainly methane. LPG (liquified petroleum gas) is a mixture of propane and butane.

The alkanes are a subset of hydrocarbons. Hydrocarbons are compounds containing hydrogen and carbon only. Examining the molecular formulae of the alkanes in table 2.5.1 shows that there is a link between the number of hydrogen atoms in the formula and the number of carbons.

📖 Word bank

- **Hydrocarbon**

A compound containing hydrogen and carbon only.

Word bank

• General formula

The formulae of compounds in a homologous series follow a pattern which can be represented by a general formula, e.g. alkanes C_nH_{2n+2} where n = 1,2,3 etc.

Number of hydrogen atoms = (number of carbon atoms $\times$ 2) + 2

This means that a **general formula** can be written that will allow you to work out the molecular formula for an alkane if you are given the number of carbons.

Alkanes have the general formula C_nH_{2n+2} where n can be 1, 2, 3, etc.

Example 2.5.1

Octane has 8 carbon atoms.

The number of hydrogens in octane would therefore be $(2 \times 8) + 2 = 18$.

Its molecular formula will therefore be C_8H_{18}.

Table 2.5.1: *The formulae for the first eight alkanes*

Alkane	Molecular formula	Full structural formula	Shortened structural formula
methane	CH_4		CH_4
ethane	C_2H_6		CH_3CH_3
propane	C_3H_8		$CH_3CH_2CH_3$
butane	C_4H_{10}		$CH_3CH_2CH_2CH_3$
pentane	C_5H_{12}		$CH_3CH_2CH_2CH_2CH_3$
hexane	C_6H_{14}		$CH_3CH_2CH_2CH_2CH_2CH_3$
heptane	C_7H_{16}		$CH_3CH_2CH_2CH_2CH_2CH_2CH_3$
octane	C_8H_{18}		$CH_3CH_2CH_2CH_2CH_2CH_2CH_2CH_3$

GO! Activity 2.5.1: Paired activity

The table shows the number of carbon atoms in four alkanes. Complete the table. For each alkane, show the number of hydrogen atoms in the molecule and the molecular formula.

Alkane	Number of carbons	Number of hydrogens	Molecular formula
nonane	9		
dodecane	12		
octadecane	18		
eicosane	20		

As well as having a general formula, members of the alkanes have similar chemical properties.

Alkanes burn in a good supply of air to give heat and carbon dioxide and water.

$$CH_4(g) + 2O_2(g) \rightarrow CO_2(g) + 2H_2O(\ell)$$

The reaction of hydrocarbon with oxygen is known as **combustion**. Combustion is an **exothermic reaction**.

When the air supply is limited combustion will be incomplete and carbon monoxide and carbon will be given off.

Carbon monoxide is particularly dangerous. Each year about 50 people die from accidental carbon monoxide poisoning. New regulations mean that whenever any appliance that might produce carbon monoxide is fitted in the home a carbon monoxide detector must also be fitted.

Carbon monoxide can cause the electrical resistance in the sensor to change or can cause a gel to become coloured. Both changes would cause the alarm to sound.

Alkanes are fairly unreactive. For a new atom to join to an alkane molecule it would need to replace a hydrogen atom that was already attached. Alkanes are said to be **saturated** compounds. This means the carbon atoms in alkanes are joined by single carbon to carbon bonds.

An example of their lack of reactivity is the reaction with bromine solution. They will only decolourise bromine solution if left standing for several minutes in direct sunlight or other strong light. The reaction takes place very slowly.

Alkanes are insoluble in water.

📖 Word bank

• **Combustion**
The reaction of a fuel with oxygen releasing energy.

📖 Word bank

• **Exothermic reaction**
A reaction in which heat energy is given out.

Figure 2.5.5: *The flame on the right shows natural gas burning in plenty of oxygen, the flame on the left is when there is less oxygen*

📖 Word bank

• **Saturated**
In saturated hydrocarbons, like alkanes, the carbon atoms are joined by single carbon to carbon bonds.

Physical properties of alkanes

Physical properties refer to properties such as melting and boiling points, volatility (how easily they change from liquid to vapour), density, etc.

Table 2.5.2: *The shortened structural formulae and boiling points of the first six straight-chain alkanes*

Alkane	Shortened structural formula	Boiling point °C
methane	CH_4	−164
ethane	CH_3CH_3	−89
propane	$CH_3CH_2CH_3$	−42
butane	$CH_3(CH_2)_2CH_3$	−1
pentane	$CH_3(CH_2)_3CH_3$	36
hexane	$CH_3(CH_2)_4CH_3$	69

The general increase in the boiling points of the alkanes is due to the increasing strength of the intermolecular forces as the size of the molecules increases. The stronger the forces, the more energy is needed to separate the molecules.

GO! Activity 2.5.2: Paired activity

1. Use the information from the table above to draw a spike graph of boiling points of alkanes against the number of carbon atoms in the molecules.
2. Discuss the boiling point trend with a partner and estimate the boiling points for heptane and octane.

📖 Word bank

• **Homologous series**

A family of compounds which have similar chemical properties and show a gradual change in physical properties (e.g. boiling point) and can be represented by the same general formula.

The physical properties of the alkanes show a regular change.

A group of compounds which have the same general formula, have similar chemical properties and show a gradation of physical properties, e.g. boiling points, is known as a **homologous series**.

The alkanes are a homologous series of saturated hydrocarbons with general formula C_nH_{2n+2}.

Naming alkanes and drawing structures

Butane has the molecular formula C_4H_{10}. Four carbons and ten hydrogens can be arranged to give two different structures.

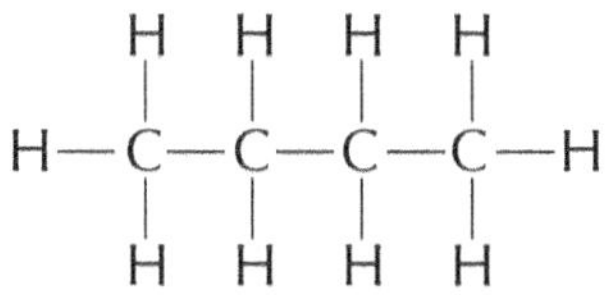

Structure 1   This arrangement is a **straight-chain** alkane.

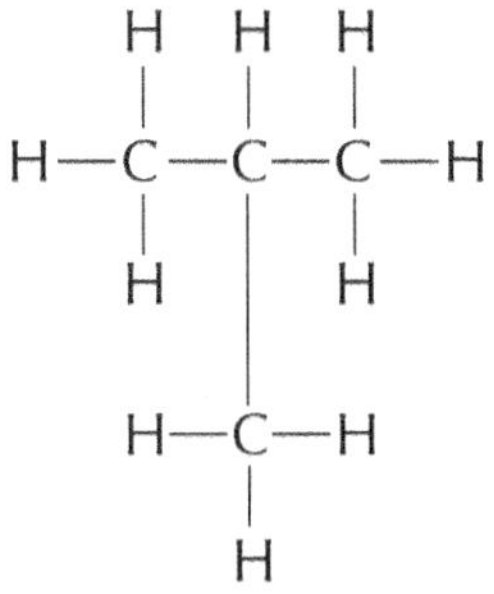

Structure 2   The second arrangement of atoms gives a **branched-chain** alkane.

Shortened structural formulae can be written for the structures above.

Structure 1 $CH_3CH_2CH_2CH_3$

Structure 2 $CH_3CH(CH_3)CH_3$

Branches are shown in brackets following the carbon they are attached to.

The molecules above are described as **isomers**. Isomers are compounds which have the same molecular formula but different structural formulae.

What's in a name?

We are all unique. Our fingerprints and our DNA identify us each as individuals. However our DNA has strong similarities to the DNA of other members of our families. We are also likely to share many characteristics. Our names reflect the family we belong to. But individually within the family we are likely to be distinguished by our first names.

In a similar way the names of organic compounds reflect the family they belong to and the first part of the name will identify the individual members.

In the alkane homologous series the names of all members will end with –ane. The individual members of the family are indicated by the first part of the name.

International rules were drawn up for the naming of organic molecules. This helped avoid confusion about which particular chemical was being referred to. The process of naming is known as **systematic naming**.

> **📖 Word bank**
>
> • **Straight-chain**
> A hydrocarbon structure consisting of carbon atoms covalently bonded to each other with no side chains.

> **📖 Word bank**
>
> • **Branched-chain**
> A hydrocarbon structure consisting of carbon atoms covalently bonded to each other with a side group attached, e.g. methyl- (–CH_3).

> **📖 Word bank**
>
> • **Isomers**
> Compounds with the same molecular formula but different structural formulae.

> **📖 Word bank**
>
> • **Systematic naming**
> An internationally agreed method of naming compounds, which is used throughout the world.

Rules for naming

For alkanes, systematic names are given by:

- naming the longest hydrocarbon chain
- naming the branches (side chains),

  e.g. CH_3— methyl, C_2H_5— ethyl, etc.

  Note: Prefixes are used if there is more than one side chain of the same type (e.g. di- is used if there are two of the same type, tri- if there are three, etc.)

- indicating the position of the branches on the main chain, numbering from the end of the main chain nearer a branch.

Example 2.5.2

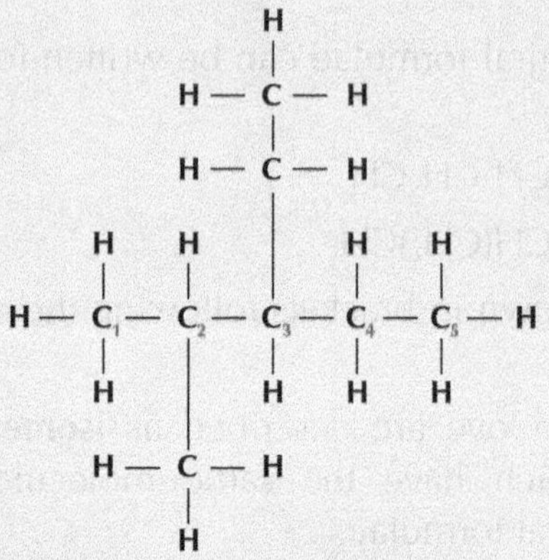

The longest chain has 5 carbons: -pentane

There are two side chains: CH_3— methyl; C_2H_5— ethyl

The ethyl is joined to the third carbon: 3-ethyl

The methyl is joined to the second carbon: 2-methyl

(Note: the side chains are arranged in alphabetical order in the name, i.e. 3-**ethyl**, 2-**methyl**.)

The name of the hydrocarbon is therefore **3-ethyl-2-methylpentane**.

When naming molecules remember the chain of carbons need not be in a straight line.

The molecule in Figure 2.5.6 is not called 2-ethylbutane because the longest carbon chain has 5 carbons. The name is therefore 3-methylpentane.

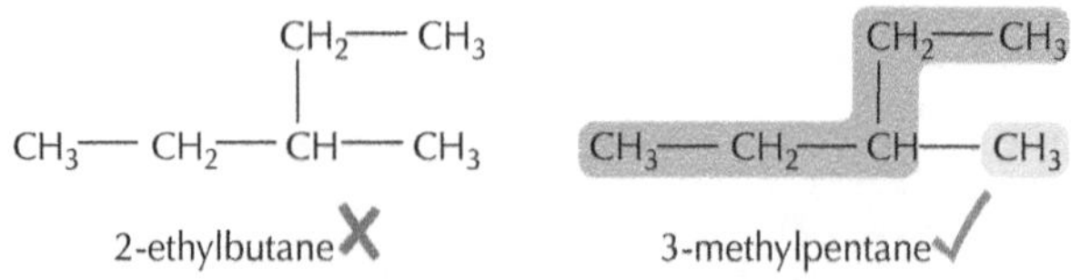

Figure 2.5.6

Drawing an isomer from its name

Drawing an alkane molecule is the reverse of naming.

- The main chain is drawn.
- The types of branches are identified.
- The positions of the branches are identified.

Draw the structural formula for 2,2,4-trimethylpentane.

Example 2.5.3

- The main chain is pentane, i.e. it has 5 carbon atoms.

C — C — C — C — C

- There are three methyl branches, i.e. 3 × CH_3.

- Two methyl groups are attached to the second carbon in the chain and one to the fourth carbon.

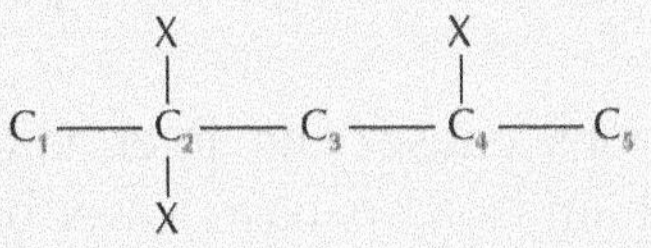

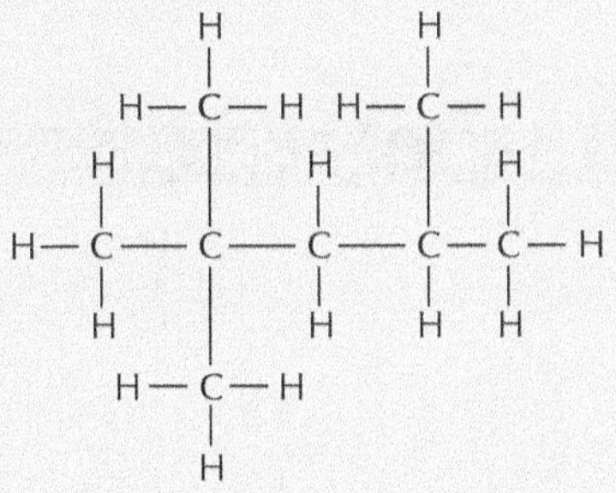

C_1 — C_2 — C_3 — C_4 — C_5 (with X branches on C_2 top and bottom, and X on C_4 top)

2,2,4-trimethylpentane

Shortened structural formula: $CH_3C(CH_3)_2CH_2CH(CH_3)CH_3$

In the shortened structural formula the branches are shown in brackets after the carbon to which they are attached.

The molecular formula can be worked out from both the structural formula and the shortened structural formula by simply counting the numbers of carbon and hydrogen atoms.

In this example the molecular formula is C_8H_{18}.

2,2,4-trimethylpentane is an isomer of octane.

GO! Activity 2.5.3: Paired activity

Working with a partner:

1. Look back at the structures of the two isomers with molecular formula C_4H_{10} on pages 99. Use the rules for naming to name Structure 2.

2. Draw full structural formulae and shortened structural formulae for the following isomers of heptane:

   3-ethylpentane
   2,4-dimethylpentane
   3,3-dimethylpentane
   2,2,3-trimethylbutane

3. Hexane, C_6H_{14}, has five isomers. Draw and name the five isomers of hexane.

Isomers don't just have different structures. Their physical properties also differ.

Branching in liquid hydrocarbons causes the intermolecular forces to decrease. Branched-chain isomers have lower boiling points and are more volatile than the straight-chain hydrocarbon.

Table 2.5.3: *The isomers of pentane*

Isomer	pentane	2-methylbutane	2,2-dimethylpropane
Structure	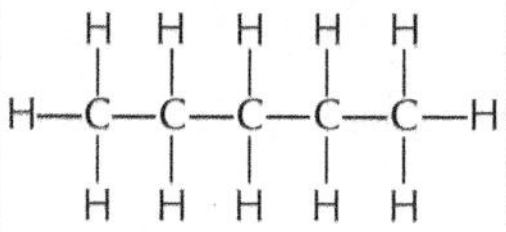	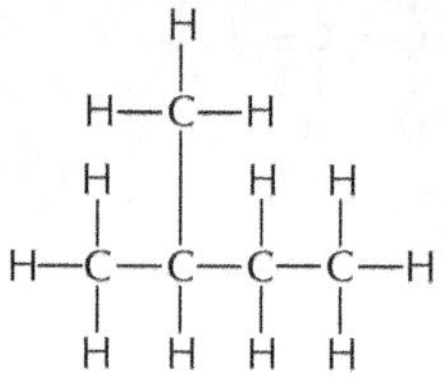	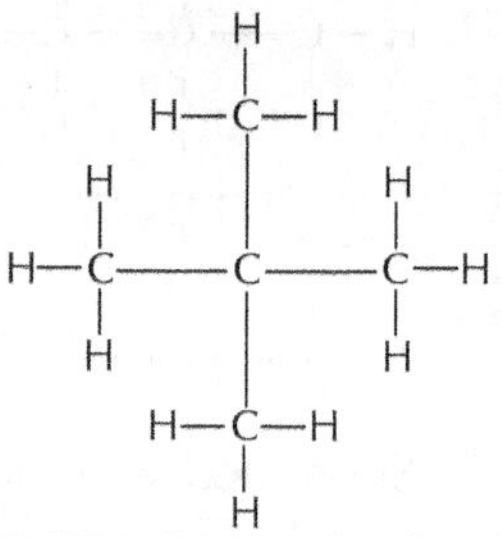
Boiling point (°C)	36	27	11

Branched-chain alkanes are particularly important in petrol manufacture, where they are used to improve the octane rating of the fuel.

Figure 2.5.7: *It may sound surprising, but Formula 1 cars use fuels very similar to ordinary petrol*

Activity 2.5.4: Paired activity

You can discuss this activity with a partner or complete it on your own.

The volatility of alkanes used in petrol is also important.

For petrol to burn in an engine it must vaporise and mix with air. The volatility of the petrol is critical. In the UK the composition of petrol is changed four times each year. In summer petrol contains less volatile components than in winter, whereas in winter, if the petrol was not volatile enough, it wouldn't vaporise and engines wouldn't start.

In what ways will the structures of molecules in a winter blend of petrol differ from those of a summer blend?

SPOTLIGHT ON PETROL AND DIESEL. WHAT'S THE DIFFERENCE?

Figure 2.5.8: *The catalytic reformer No.3 at the Kwinana refinery in Australia © BP plc*

Both petrol and diesel are mixtures of hydrocarbons. Petrol and diesel differ in their composition. Petrol is made from the naphtha fraction of crude oil, which is lighter than the diesel fraction. Petrol will therefore contain smaller molecules. Petrol will also contain more branched-chain hydrocarbons. Branched-chain alkanes can be produced by a process known as reforming. Straight-chain alkanes are passed over a heated platinum catalyst. This breaks the hydrocarbon chain. When the fragments are allowed to reform, branched-chain alkanes can be formed. In reforming, the number of carbon atoms in the molecules always remains the same but the atoms are joined together in different ways forming new compounds. When straight-chain alkanes are

Figure 2.5.9: *A car mechanic checks for engine damage*

converted to branched-chain isomers the reforming process is sometimes referred to as isomerisation.

In petrol engines a fuel/air mixture is compressed and ignited by a spark. 'Engine knocking' in petrol engines is caused by the fuel/air mixture igniting spontaneously as it is compressed. This can cause damage to the engine and leads to a loss of efficiency.

Petrol is given an octane rating. This is a measure of its tendency to auto-ignite and cause knocking. The petrol is compared to a mixture of 2,2,4-trimethylpentane and heptane.

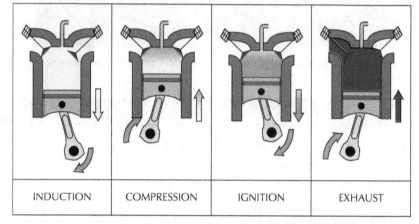

| INDUCTION | COMPRESSION | IGNITION | EXHAUST |

Figure 2.5.10: *The cycle of strokes in a four-stroke diesel engine*

2,2,4-trimethylpentane has been assigned an octane rating of 100. It has low tendency to auto-ignite and cause knocking. Heptane has been assigned an octane rating of 0. It tends to auto-ignite, causing severe engine knocking. 95 octane petrol has the same knocking characteristic as a mixture of 95% 2,2,4-trimethylpentane and 5% heptane.

Engines that burn diesel are different from those that burn petrol. In diesel engines there is no spark. Air is compressed in the engine cylinder. This causes the air to heat up. Diesel fuel is then injected into the engine cylinder and spontaneously ignites.

- Induction: air is drawn into the cylinder.
- Compression: air is compressed, causing it to heat up.
- Ignition: diesel fuel is injected, igniting spontaneously.
- Exhaust: exhaust gases are pushed out of the cylinder.

Diesel fuels have a low octane rating. The components tend to auto-ignite. The industry gives diesel a cetane rating. Cetane is a mixture of straight-chain alkanes that ignites very easily when compressed.

Cycloalkanes

As the name suggests, **cycloalkanes** are hydrocarbon compounds with the carbons arranged in a ring.

Like the alkanes, these molecules are saturated, i.e. the carbons are joined to each other by single carbon to carbon bonds. The cycloalkanes are named simply by putting the prefix cyclo- in front of the name of the straight-chain alkane with the same number of carbon atoms.

Cylcoalkanes react in a similar way to alkanes with the exception of cyclopropane, which is more reactive than might be expected. This is because in cyclopropane the angles between the carbon to carbon bonds are only 60°, creating great strain in the ring.

Counting carbon and hydrogen atoms gives C_nH_{2n} as the general formula for the cycloalkanes, where n = 3, 4, 5 etc.

Cyclohexane occurs naturally in crude oil and can be obtained by fractional distillation. It is also known to be present in volcanic emissions.

> 📖 **Word bank**
>
> • **Cycloalkanes**
> A homologous series of saturated hydrocarbon with the carbons arranged in a ring.

Table 2.5.4: *The first four members of the cycloalkane homologous series*

Name	Molecular formula	Structural formula	Shortened structural formula
cyclopropane	C_3H_6	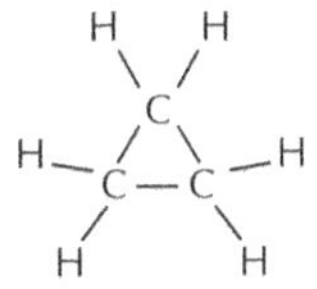	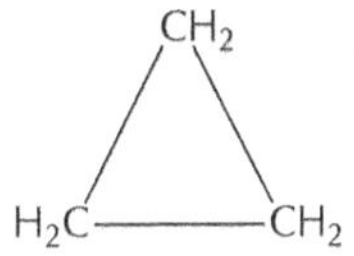
cyclobutane	C_4H_8	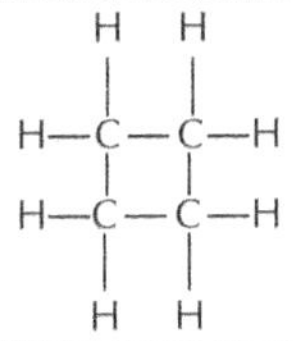	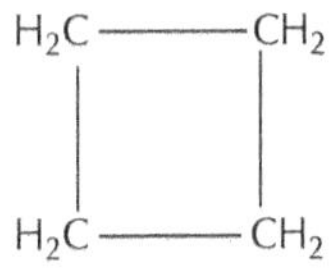
cyclopentane	C_5H_{10}	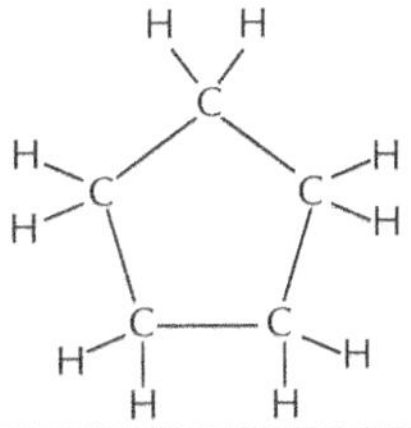	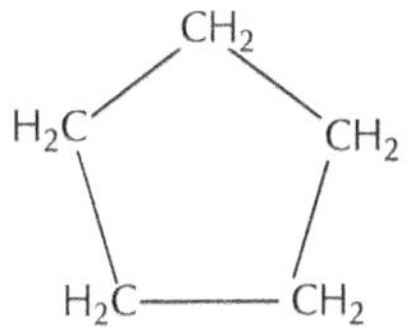
cyclohexane	C_6H_{12}	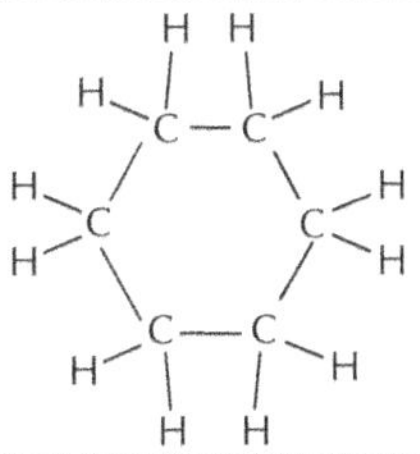	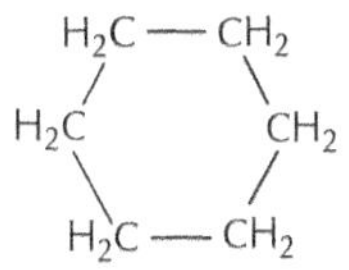

Figure 2.5.11: *Cyclopentane is now being used to create the foam for insulating refrigerators*

Uses of cycloalkanes

Cycloalkanes are used in motor fuels.

Cyclopentane is replacing chlorofluorocarbons (CFCs) as a blowing agent when creating polyurethane foams for insulation in domestic appliances such as refrigerators.

Unlike CFCs, cyclopentane does not cause ozone depletion.

The principal use of cyclohexane is in the manufacture of the chemicals adipic acid and caprolactam, which are used in the manufacture of nylon.

Cycloheptane is used as a solvent for the chemical industry and as an intermediate in the manufacture of chemicals and pharmaceutical drugs.

GO! Activity 2.5.5

1. (a) Name the cycloalkane with formula C_7H_{14}.
   (b) Draw the structural formula for the cycloalkane in part (a).
   (c) Name an alkene isomer of the cycloalkane in part (a).

2. (a) Name the cycloalkane with the following structure:
   (b) Write the molecular formula for the cycloalkane in part (a).
   (c) Oct-1-ene has the same molecular formula but a different structural formula to the cycloalkane in part (a). Give the term used to describe this.

Alkenes

The alkenes are a homologous series of hydrocarbons that contain a double carbon to carbon bond between two of the carbon atoms in the structure. The structural formula of ethene (the first alkene) and propene are shown. Because of the C = C, double-bond alkenes are said to be **unsaturated**.

ethene

propene

The general formula for the alkenes is C_nH_{2n} where n = 2, 3, 4 etc.

Butene has molecular formula C_4H_8. There are different alkene isomers with molecular formula C_4H_8. The straight-chain isomers differ in the position of the carbon to carbon double bond within the molecule. There is also a branched-chain isomer.

The double carbon to carbon bond is attached to the first carbon in the chain.

The double carbon to carbon bond is between the second and third carbons in the chain.

The simplest branched-chain alkene.

Every member of the alkene homologous series from butene onwards has isomers both due to the positioning of the double bond and due to branching.

Since alkenes and cycloalkanes have the same general formula, alkenes and cycloalkanes with the same number of carbon atoms are isomers.

Naming and drawing alkenes

For alkenes, the name identifies the position of the double bond within the molecule. The rules for naming are therefore slightly different from naming alkanes. The name is worked out by:

- naming the longest hydrocarbon chain containing the double bond
- numbering the chain from the end nearer the double bond
- naming the branches (side chains). Remember: prefixes are used if there is more than one side chain of the same type (e.g. di- is used if there are two of the same type, tri- if there are three, etc.)
- indicating the position of the side chains on the main chain.

Example 2.5.4

Naming alkenes

The structure below represents an isomer of hexene:

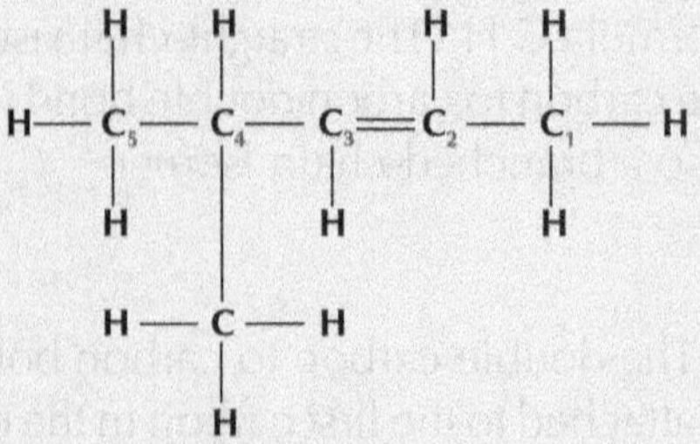

Naming the compound:

The longest chain has 5 carbons: pentene

The double bond starts on the second carbon from the right and the third carbon from the left. The chain is therefore numbered from the right:

pent-2-ene

(Note: the number indicating the carbon to carbon double bond position goes in the middle of the name.)

There is one side chain: CH_3— methyl

The methyl is joined to the fourth carbon: 4-methyl

The name of the alkene is therefore: 4-methylpent-2-ene

Example 2.5.5

Drawing alkenes

Draw the structure for **3,4-dimethylpent-2-ene**.

Pentene: the longest chain has 5 carbon atoms.

C–C–C–C–C

Pent-**2**-ene: There is a double carbon to carbon bond between the second and third carbons in the chain.

5 4 3 2 1
C–C–C=C–C

Dimethylpent-2-ene: There are two methyl (CH_3) branches.

3,4-dimethylpent-2-ene: The methyl branches are attached to the third and fourth carbon atoms of the chain.

```
        X
        |
C–C–C=C–C
    |
    X
```

🔍 **Hint**

Drawing an alkene molecule is the reverse of naming.

- Draw the main chain including the double bond
- Number the chain from the end nearest the double bond
- Identify the types of branches
- Identify the position of the branches

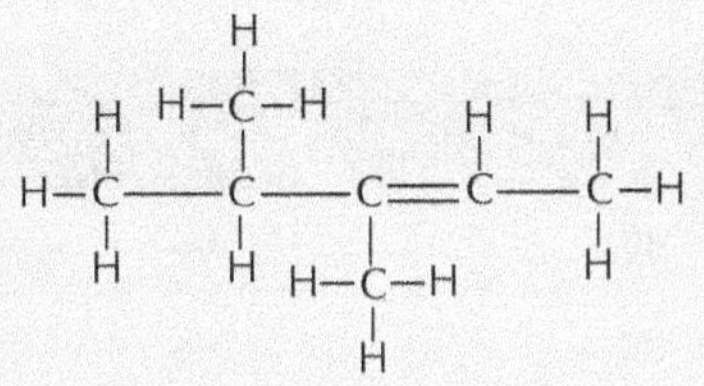

3,4-dimethylpent-2-ene

Shortened structural formula: $CH_3CH(CH_3)C(CH_3)CHCH_3$

The molecular formula can be worked out from the full structural formula or the shortened structural formulae.

Molecular formula: C_7H_{14}

The general formula for the alkenes is the same as that for cycloalkanes. Cyclopropane would therefore be an isomer of propene, the alkene with three carbon atoms.

3,4-dimethylpent-2-ene has molecular formula C_7H_{14} and would therefore be an isomer of cycloheptane.

⊙ Activity 2.5.6: Paired activity

Using the rules for naming alkenes.

1. Name the alkene isomers of butene C_4H_8 (page 107).

2.

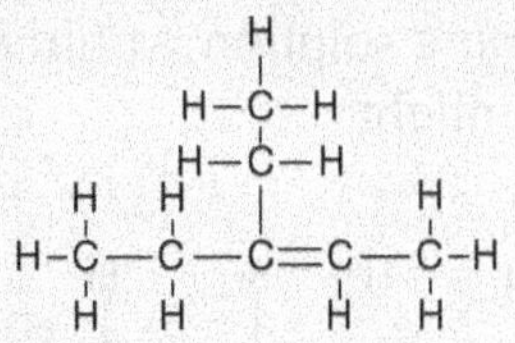

   (a) Name the alkene drawn above.
   (b) Write its shortened structural formula and molecular formula.

3. Draw and name as many of the isomers of hexene as you can. There are three straight-chain isomers and ten branched-chain isomers. If you have access to a molymod kit, it might help to build models to help you name the isomers.

> **🔎 Hint**
>
> Can't get the tenth branched-chain isomer? Do you have the branched-chain isomer that has an ethyl side chain?

Physical properties

Like the alkanes, alkene isomers show differences in their physical properties. Again this is due to the strength of the forces between the molecules of the isomers. The bigger the molecules, the larger the forces of attraction between them, so more energy is needed to separate them. This results in higher melting and boiling points.

Isomers of hexene have different melting and boiling points.

Table 2.5.5: *Isomers of hexene showing differences in melting and boiling points*

Isomer	Melting point (°C)	Boiling point (°C)
hex-1-ene	−140	63
hex-2-ene	− 99	68
hex-3-ene	−113	67
4-methylpent-1-ene	−153	53

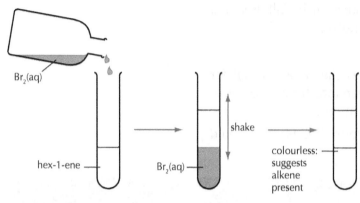

Figure 2.5.12: *Bromine solution is immediately decolourised when shaken with hex-1-ene*

Reactivity of alkenes

Alkenes are very reactive due to the presence of the carbon to carbon double bond in the molecules. They are therefore important feedstocks in the chemical industry, being used to produce many other important chemicals.

Alkenes can be distinguished from alkanes using bromine solution.

When bromine solution is added to hex-1-ene the bromine solution is decolourised immediately. If bromine solution was added to the alkane, hexane, the bromine solution would only decolourise if left standing in strong sunlight.

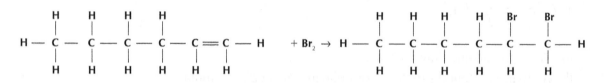

hex-1-ene (colourless)   bromine solution (orange)   1,2-dibromohexane (colourless)

📖 Word bank

• **Addition reaction**

One of the bonds of the carbon to carbon double bond is broken and new atoms (or groups) join to the carbon chain.

The reaction with bromine is an example of an **addition** reaction. In an addition reaction one of the bonds of the carbon to carbon double bond is broken and new atoms (or groups) join to the carbon chain.

Alkenes are said to be unsaturated because new atoms or groups can be added into the molecule. The product molecules formed when alkenes react by addition are said to be saturated. The carbons in the molecules are joined by single carbon to carbon bonds.

Other addition reactions

Adding other halogens

As well as bromine other halogens are able to add to alkenes, forming a dihaloalkane.

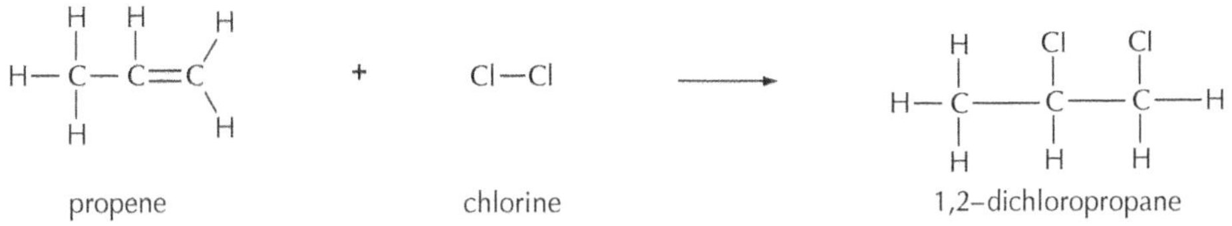
propene
chlorine
1,2-dichloropropane

Adding hydrogen

Hydrogen can be added across the double carbon to carbon bond, giving the corresponding alkane. The process is also known as **hydrogenation**.

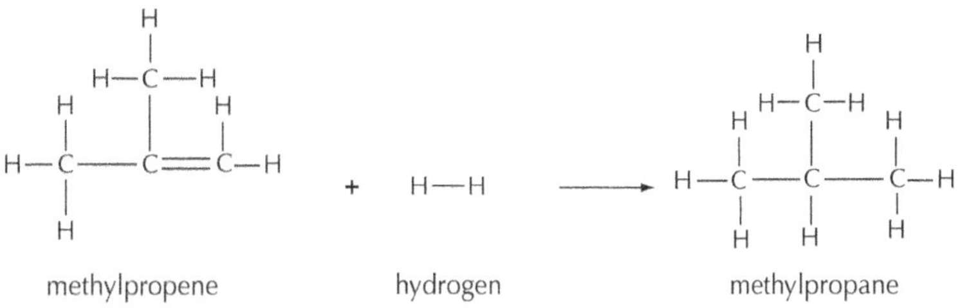
methylpropene
hydrogen
methylpropane

Word bank

- **Hydrogenation**

An addition reaction where hydrogen adds on to a molecule across the double carbon to carbon bond in an alkene (unsaturated), giving the corresponding alkane (saturated).

SPOTLIGHT ON THE FOOD INDUSTRY

Adding hydrogen across double carbon to carbon bonds is an important process in margarine manufacture.

Margarine and butter substitutes are made by mixing vegetable oils and skimmed milk to give an emulsion. Oils are liquid at room temperature due to the molecules in oil containing hydrocarbon chains with carbon to carbon double bonds between some of the carbon atoms in the chain. The melting point of the oil can be raised by removing some of the carbon to carbon double bonds. This is done by heating the oil and bubbling hydrogen through the oil in the presence of a nickel catalyst. The process is known as hydrogenation or hardening.

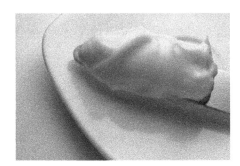

Figure 2.5.13: *During margarine manufacture 'hardened' oils and skimmed milk are mixed to form an emulsion*

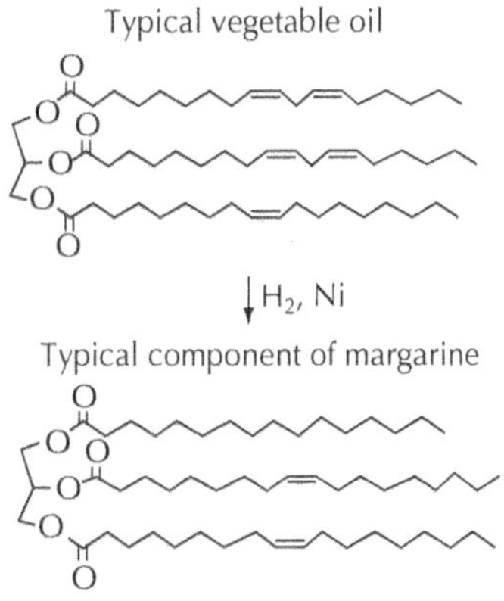
Typical vegetable oil
H2, Ni
Typical component of margarine

Figure 2.5.14: *Partial hydrogenation of a typical plant oil to a typical component of margarine*

Adding water

Although alkanes are insoluble in water, they can be made to react with it. The addition reaction, which involves adding water across a double carbon to carbon bond, is also referred to as **hydration**.

In industry this is achieved by reacting the alkene with steam at 300 °C and 70 atmospheres pressure using phosphoric acid as a catalyst.

H—C=C—H   +   H—OH   ⟶   H—C————C—H (with H and OH, H on carbons)

ethene   +   steam   ⟶   ethanol

Propanol and butanol, two important industrial solvents, are also made by the hydration of alkenes.

propene   +   steam   ⟶   propanol

When hydrogen compounds are added to alkenes, the hydrogen atom tends to bond to the carbon which has most Hs already attached.

When water is added to propene, a hydrogen attaches to the end carbon and a hydroxyl group adds to the middle carbon.

Making plastics

The most important use of alkenes and alkene derivatives is as feedstocks in the plastics industry. This is dealt with fully in Chapter 8: Plastics.

Learning checklist

In this chapter you have learned:

- The alkanes are a homologous series of hydrocarbons with general formula C_nH_{2n+2}.
- The members of the alkane homologous series show similar chemical properties and show a gradation in their physical properties such as boiling points.
- Alkanes are generally unreactive but burn and are used as fuels.
- The alkanes are saturated hydrocarbons with only single carbon to carbon bonds between carbon atoms.
- To name branched-chain alkanes given their structure and draw structural formulae given the systematic name of an alkane.
- Cycloalkanes are a homologous series of saturated hydrocarbons where the carbons are joined in a ring.
- Some uses for cycloalkanes.
- To name cycloalkanes, given their formulae.
- Alkenes are a homologous series of hydrocarbons with general formula C_nH_{2n}.
- Alkenes are unsaturated – they contain a C=C double bond.
- To systematically name straight-chain and branched-chain-alkenes and draw structural formulae given the systematic name of the alkene.
- Alkenes contain a double carbon to carbon bond, which makes them more reactive than alkanes.
- Alkenes can undergo addition reactions when other molecules join by adding across the double bond.
- Hydrogenation and hydration are two examples of addition reactions with important industrial applications.
- Isomers have the same molecular formula but different structural formulae.
- How to draw structural formulae for isomers of alkanes, alkenes and cycloalkanes.
- Alkanes, alkenes amd cycloalkanes are all insoluble in water.

6 Everyday consumer products

You should already know

- Ethanol is a fuel made from plants.

Learning intentions

In this chapter you will learn about:

- Alcohols – naming and drawing alcohols, alcohols as fuels, alcohols as solvents.
- Energy from fuels – exothermic and endothermic reactions, comparing fuels and calculating mass of reactants and products.
- Carboxylic acids – naming and drawing, and everyday uses.

Alcohols

Alcohol is the popular name for a particular alcohol, ethanol, which is found in alcoholic drinks such as whisky.

Figure 2.6.1: *Pot stills used in whisky production*

▬◀ SPOTLIGHT ON THE WHISKY INDUSTRY

Whisky is perhaps Scotland's best known export. The name comes from the Gaelic 'Uisge Beatha' which means 'water of life'. Whisky is worth more than £2 billion to the economy. The ethanol is produced by breaking down the starches in crops such as barley, fermenting the sugars produced and then distilling the fermented liquor in copper stills to produce the raw spirit. The raw spirit is a mixture of ethanol and water. It has a high percentage of ethanol as ethanol boils at 78°C.

The spirit must then be matured in wooden barrels for at least three years to become whisky. At any one time there are as many as 18½ million barrels of whisky maturing in warehouses throughout Scotland. Every year 2% of the whisky evaporates into the air. This is known as the angels' share.

Ethanol is the second member of a homologous series of **alcohols**, sometimes known as alkanols, based on the alkanes.

$$H - \overset{\overset{\displaystyle H}{|}}{\underset{\underset{\displaystyle H}{|}}{C}} - \overset{\overset{\displaystyle H}{|}}{\underset{\underset{\displaystyle H}{|}}{C}} - O - H$$

ethanol

One method of making alcohols is by the hydration of alkenes (see Area 2, Chapter 5: Homologous series), e.g. ethanol can be made by the hydration of ethene. This is carried out industrially by reacting ethene with steam in the presence of a phosphoric acid catalyst at 300°C and 70 atmospheres pressure.

Activity 2.6.1

Draw structural formulae for the alcohols that can be made by the hydration of propene and pent-1-ene.

The names and formulae of the alcohols are related to the names and formulae of the corresponding alkanes. In the alcohols a hydrogen in the alkane molecule has been replaced by an oxygen bonded to a hydrogen. This is known as a hydroxyl group. All alcohols contain a hydroxyl group (-OH). This is the **functional group** for alcohols and is responsible for the similarities in properties of the members of the homologous series. The 'e' from the end of the alkane name is replaced with 'ol' to give the name of the alcohol.

Table 2.6.1: *Molecular and structural formulae for the first three members of the alcohol homologous series based on alkanes*

Alkane	Molecular formula	Alcohol	Molecular formula	Structural formula
methane	CH_4	methanol	CH_3OH	$H-\overset{\overset{H}{\vert}}{\underset{\underset{H}{\vert}}{C}}-OH$
ethane	C_2H_6	ethanol	C_2H_5OH	$H-\overset{\overset{H}{\vert}}{\underset{\underset{H}{\vert}}{C}}-\overset{\overset{H}{\vert}}{\underset{\underset{H}{\vert}}{C}}-OH$
propane	C_3H_8	propanol	C_3H_7OH	$H-\overset{\overset{H}{\vert}}{\underset{\underset{H}{\vert}}{C}}-\overset{\overset{H}{\vert}}{\underset{\underset{H}{\vert}}{C}}-\overset{\overset{H}{\vert}}{\underset{\underset{H}{\vert}}{C}}-OH$

From the information in the table the general formula for the alcohols can be worked out. The general formula for the alcohols based on alkanes is $C_nH_{2n+1}OH$, where n = 1,2,3 etc.

Alcohol isomers

Two different alcohols with formula C_3H_7OH exist. The hydroxyl group in these isomers is attached to the carbon chain at different places.

Full structural formulae

All alcohols with more than two carbons have isomers due to the position of the hydroxyl group in the molecule.

Shortened structural formulae

$CH_3CH_2CH_2OH$

The hydroxyl functional group is attached to an end carbon.

$CH_3CH(OH)CH_3$

The hydroxyl functional group is attached to the middle carbon in the chain.

All alcohols with more than two carbons have isomers due to the position of the hydroxyl group in the molecule.

Naming and drawing alcohols

The name of an alcohol specifies where the hydroxyl group is attached to the carbon chain. The rules for naming are similar to those for naming alkenes (see Chapter 5: Naming and drawing alkenes).

- Name the longest hydrocarbon chain that has a hydroxyl group attached.

- Number the chain from the end nearer the hydroxyl group. (Note: You are only required to be able to name straight-chain alcohol isomers.)

  The next two rules are the same as for naming alkenes and would allow you to name an alcohol that had a branched hydrocarbon chain.

- Name the branches (side chains). Remember: prefixes are used if there is more than one side chain of the same type (e.g. di- is used if there are two of the same type, tri- if there are three, etc.).

- Indicate, using numbers, the position of the side chains on the main chain.

Naming the alcohol: **Example 2.6.1**

$$H-\overset{\overset{\displaystyle H}{|}}{\underset{\underset{\displaystyle H}{|}}{C}}-\overset{\overset{\displaystyle H}{|}}{\underset{\underset{\displaystyle H}{|}}{C}}-\overset{\overset{\displaystyle H}{|}}{\underset{\underset{\displaystyle H}{|}}{C}}-\overset{\overset{\displaystyle H}{|}}{\underset{\underset{\displaystyle O}{|}}{C}}-\overset{\overset{\displaystyle H}{|}}{\underset{\underset{\displaystyle H}{|}}{C}}-\overset{\overset{\displaystyle H}{|}}{\underset{\underset{\displaystyle H}{|}}{C}}-H$$

- The carbon chain contains 6 carbons. The alcohol is an isomer of hexanol.

- The chain is numbered from the right since this is the end nearer the hydroxyl group. The hydroxyl is on the third carbon. The name of the alcohol would therefore be **hexan-3-ol**.

Drawing the alcohol:

Draw the full structural formula for heptan-3-ol and write a shortened structural formula for the compound.

- The alcohol has 7 carbons: heptan = 7.

- The hydroxyl group is attached to the third carbon in the chain. (It doesn't matter which end you start from.)

$$C_7-C_6-C_5-C_4-\underset{\underset{\displaystyle X}{|}}{C_3}-C_2-C_1$$

Full structural formula:

$$H-\overset{\overset{\displaystyle H}{|}}{\underset{\underset{\displaystyle H}{|}}{C}}-\overset{\overset{\displaystyle H}{|}}{\underset{\underset{\displaystyle H}{|}}{C}}-\overset{\overset{\displaystyle H}{|}}{\underset{\underset{\displaystyle H}{|}}{C}}-\overset{\overset{\displaystyle H}{|}}{\underset{\underset{\displaystyle O}{|}}{C}}-\overset{\overset{\displaystyle H}{|}}{\underset{\underset{\displaystyle H}{|}}{C}}-\overset{\overset{\displaystyle H}{|}}{\underset{\underset{\displaystyle H}{|}}{C}}-\overset{\overset{\displaystyle H}{|}}{\underset{\underset{\displaystyle H}{|}}{C}}-H$$

Shortened structural formula:

$$CH_3CH_2CH_2CH_2CH(OH)CH_2CH_3$$

The molecular formula can be worked out from the structural and shortened structural formulae: $C_7H_{15}OH$.

Physical properties of alcohols

The alcohols are a homologous series, so show a gradual change in physical properties.

Table 2.6.2 shows the trend in melting and boiling points for alcohols with the hydroxyl group attached to an end carbon of the chain.

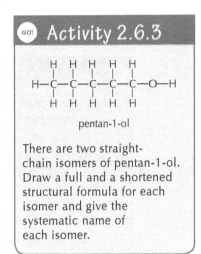

GO! Activity 2.6.3

$$H-\overset{\overset{\displaystyle H}{|}}{\underset{\underset{\displaystyle H}{|}}{C}}-\overset{\overset{\displaystyle H}{|}}{\underset{\underset{\displaystyle H}{|}}{C}}-\overset{\overset{\displaystyle H}{|}}{\underset{\underset{\displaystyle H}{|}}{C}}-\overset{\overset{\displaystyle H}{|}}{\underset{\underset{\displaystyle H}{|}}{C}}-\overset{\overset{\displaystyle H}{|}}{\underset{\underset{\displaystyle H}{|}}{C}}-O-H$$

pentan-1-ol

There are two straight-chain isomers of pentan-1-ol. Draw a full and a shortened structural formula for each isomer and give the systematic name of each isomer.

Table 2.6.2: *Melting and boiling points of the straight-chain alcohols*

Alcohol	Shortened structural formula	Melting point/°C	Boiling point/°C
methanol	CH_3OH	−98	65
ethanol	CH_3CH_2OH	−117	78
propan-1-ol	$CH_3\,CH_2CH_2OH$	−127	97
butan-1-ol	$CH_3(CH_2)_2CH_2OH$	−90	116
pentan-1-ol	$CH_3(CH_2)_3CH_2OH$	−79	137
hexan-1-ol	$CH_3(CH_2)_4CH_2OH$	(a)	(d)
heptan-1-ol	$CH_3(CH_2)_5CH_2OH$	(b)	177
octan-1-ol	$CH_3(CH_2)_6CH_2OH$	(c)	194

GO! Activity 2.6.4

1. (a) Use information from the spike graph of the melting points against number of carbon atoms to create a table for the melting points of hexan-1-ol, heptan-1-ol and octan-1-ol.

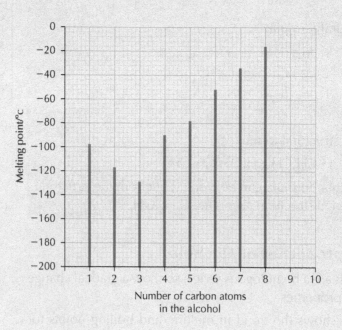

(b) (i) Draw a spike graph of the boiling points of the alcohols in Table 2.6.2 against the number of carbons.

  (ii) Use your graph to estimate the boiling point of hexan-1-ol.

Explaining the trend in boiling points

The boiling points of the alcohols increase as the number of carbon atoms in the molecules increase. Larger molecules with longer hydrocarbon chains lead to greater forces of attraction between molecules. This means more energy will be required to separate heavier molecules and boiling points will increase.

Figure 2.6.2: *Strong intermolecular attractions in methanol*

Comparing the boiling point of an alcohol with that of an alkane of similar molecular mass shows that there are stronger intermolecular forces between alcohol molecules. This is due to the presence of the hydroxyl group in alcohol molecules. The forces of attraction between hydroxyl groups are much greater than the forces of attraction between hydrocarbon parts of molecules.

Propane, C_3H_8
(Formula mass: 44)
mp −188°C; bp −44°C

Ethanol, C_2H_5OH
(Formula mass: 46)
mp −117°C; bp 78°C

Isomeric alcohols show differences in boiling points. When the hydroxyl group is attached to an end carbon the forces of attraction between molecules is stronger. Alcohols with the hydroxyl group attached to an end carbon therefore have higher boiling points than a corresponding isomeric alcohol with the hydroxyl group attached to the middle of the chain.

Table 2.6.3

Alcohols with hydroxyl on an end carbon	Boiling point/°C	Alcohols with the hydroxyl group on a middle carbon	Boiling point/°C
propan-1-ol	97	propan-2-ol	82
butan-1-ol	116	butan-2-ol	100
pentan-1-ol	137	pentan-2-ol	118

Solubility in water

Methanol and ethanol are totally miscible with water, that is, they mix completely with water in all proportions. Strong intermolecular forces can form between the small alcohol molecules and the water molecules. The forces of attraction between the hydrocarbon parts of larger alcohol molecules make it more difficult for these alcohol molecules to separate and dissolve in water.

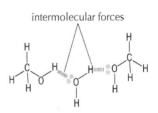

Figure 2.6.3: *Strong intermolecular forces between methanol and water molecules*

As the hydrocarbon chain length of the alcohols in the series increases, the solubility of the alcohols decreases.

Table 2.6.4

Alcohol	Formula	Solubility in water/g l⁻¹
butan-1-ol	C_4H_9OH	63
pentan-1-ol	$C_5H_{11}OH$	22
hexan-1-ol	$C_6H_{13}OH$	5·9
heptan-1-ol	$C_7H_{15}OH$	1·7
octan-1-ol	$C_8H_{17}OH$	0·5

Figure 2.6.4: *The extract from vanilla pods is dissolved in ethanol*

Alcohols as solvents

Ethanol is an effective solvent. After water it is the most important solvent used by industry.

The shapes of methanol and ethanol molecules are similar to that of water. Like water, methanol and ethanol are good solvents.

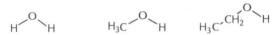

Figure 2.6.5: *The shapes of methanol and ethanol molecules are similar to that of water*

Ethanol, as well as being used extensively in industry, is a common solvent in perfumes, food flavourings and in many medicinal preparations.

Propan-2-ol is also widely used as a solvent. It dissolves a wide range of compounds and is used in cleaning fluids to dissolve oils. It is used to clean computer keyboards and monitor screens. It evaporates quickly and is relatively non-toxic, compared to alternative solvents.

Propan-2-ol is also the alcohol used in many hand gels and disinfectant wipes.

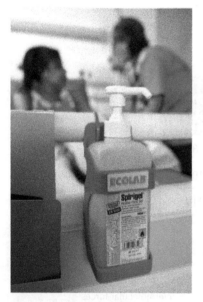

Figure 2.6.6: *Alcohol-based hand gels are currently used in the fight against the hospital superbug MRSA*

Energy from fuels

Alcohols are very flammable and can be used as fuels. Methanol and ethanol burn cleanly to give carbon dioxide and water.

$$C_2H_5OH(\ell) \ + \ 3O_2(g) \ \rightarrow \ 2CO_2(g) \ + \ 3H_2O(\ell)$$

ethanol

Figure 2.6.7: *Methylated spirit (meths), a mixture of the alcohols ethanol and methanol, can be used as fuel in many camping stoves*

Activity 2.6.5

Write balanced equations for the complete combustion of:

(a) methanol

(b) propan-1-ol.

Hint

Be careful when balancing the equation – remember there is an oxygen in the alcohol molecule!

Unlike hydrocarbons, methanol and ethanol do not produce carbon monoxide when they burn. Both methanol and ethanol can be used as fuels in car engines. Alcohols with more carbons in the hydrocarbon chain are more likely to burn incompletely and can release soot and carbon monoxide.

SPOTLIGHT ON CAR ENGINES

There is increasing use of ethanol as a fuel for cars. Normally the ethanol is blended with petrol to give a petrol/ethanol mixture. E10 fuel would be a mixture of 90% normal petrol and 10% ethanol. In the US and Europe cars have been produced which will run on an E85 fuel mixture (15% petrol; 85% ethanol).

In Brazil, engine developments allow cars to run on E100 fuel, i.e. pure ethanol. Although ethanol fuels do not give the same mileage and more needs to be used, the cost of fuel and the greenhouse gas emissions are lower.

Ethanol for fuels is being produced by fermenting corn. Ethanol is therefore a renewable source whereas the fossil fuels used to produce petrol are a finite resource.

Modifying engines to burn fuels containing ethanol can be done cheaply and easily. As petrol from fossil fuels becomes more expensive it will become more economical to use ethanol as a fuel. Since 1998 many cars manufactured in the United States have been able to run on either petrol or E85.

Figure 2.6.8: *The cost of alcohol fuel is lower than petrol and diesel*

Table 2.6.5: *Comparison of a car's performance using petrol and E85 as fuel*

Petrol:	E85:
MPG (city): 21 (11·2 l/100 km)	*MPG (city): 16 (14·7 l/100 km)*
MPG (highway): 31 (7·6 l/100 km)	*MPG (highway): 24 (9·8 l/100 km)*
Greenhouse Gas Emissions: 7·4	*Greenhouse Gas Emissions: 5·9*

Source: http://interestingenergyfacts.blogspot.co.uk

Methanol can also be used as a fuel and is used in drag racing.

Some TM Dragsters burn methanol in supercharged engines – they can cover the quarter-mile drag strip in 5·3 seconds at over 270 mph.

Figure 2.6.9: *Dragsters ready to race*

GO! Activity 2.6.6: Paired activity

Discuss advantages and disadvantages of using ethanol instead of petrol as a fuel. You may wish to use the internet to gather more information on this.

Comparing different fuels

Why do motorists who drive cars with diesel engines tend to get better mileage than those using cars with petrol engines?

Although diesel-engine cars are more expensive and diesel tends to be more expensive than petrol, fuel efficiency can make them more economical for high-mileage driving.

The reason is partly to do with the energy density of the fuels. The energy density of the fuel is the energy given out when a unit mass (e.g. a kilogram) or a unit volume (e.g. a litre) of a fuel is burned.

Diesel has an energy density that is about 13% greater than petrol. The table below shows the energy densities of various fuels in comparison to petrol. The higher the energy density, the more energy is produced, and the further a vehicle should travel per litre of fuel used.

Table 2.6.6: *The energy densities of different fuels*

Fuel	Energy density (relative to petrol)/%
petrol	100
diesel	113
biodiesel	103
E10 (90% petrol; 10% ethanol)	93
E85 (15% petrol; 85% ethanol)	73

Calorimeters

The energy given out when a fuel burns can be measured using a calorimeter. Sophisticated calorimeters can give very accurate measurements of the energy produced.

In an oxygen bomb calorimeter the fuel sample is placed in a container filled with oxygen. The fuel is ignited using an electrical ignition coil. The energy released by the burning fuel heats water, which is being constantly stirred to ensure that all of the water is at the same temperature. Accurate measurement of the temperature of the water allows the heat produced by the burning fuel to be calculated.

> ### ⚗️ Make the link
>
> The word 'calorie' is more commonly used in reference to food and you would cover it on a Health & Food Technology course. As a measurement of energy, the word is relevant to Chemistry too.

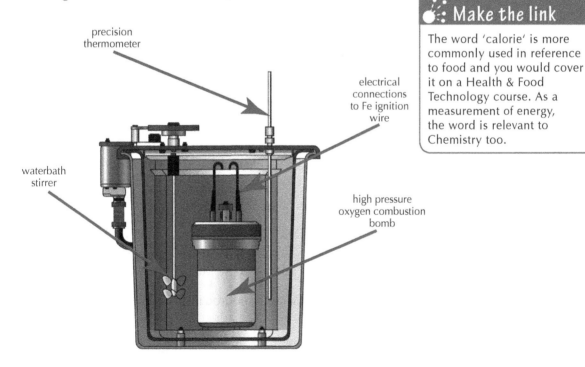

precision thermometer

electrical connections to Fe ignition wire

waterbath stirrer

high pressure oxygen combustion bomb

Parr Model 1341 Oxygen Bomb Calorimeter

Figure 2.6.10: *Bomb calorimeters are used to measure the energy given from fuels*

Calculating the energy transferred

4·18 kilojoules of energy are required to raise the temperature of 1 kilogram of water by 1°C. This is known as the specific heat capacity of water. The symbol for this is 'c'.

Specific heat capacity of water (c) = $4·18 \text{ kJ kg}^{-1} \text{ °C}^{-1}$

The energy required to heat a mass of water can therefore be calculated by multiplying the specific heat capacity of water by the mass of water being heated and the temperature rise, i.e.

energy required = **specific heat capacity of water** × **mass of water** × **temperature rise**

$$E_h = c \times m \times \Delta T$$

> ### 🔍 Hint
> The relationship $E_h = cm\Delta T$ and the value of c, can be found in the SQA data booklet.

Example 2.6.2

Calculate the energy required to raise the temperature of 500 cm³ of water by 12 °C.

$c = 4·18 \text{ kJ kg}^{-1} \text{ °C}^{-1}$

$m = 0·5 \text{ kg}$ (500 cm³ = 500 g = 0·5 kg)

(remember the mass must be changed to kilograms)

$\Delta T = 12°C$

E_h (energy required) = $cm\Delta T$

$= 4·18 \times 0·5 \times 12 \text{ kJ}$
$(\text{kg}^{-1} \text{ °C}^{-1} \times \text{kg} \times \text{°C} = \text{kJ})$

$$\mathbf{E_h = 25·08 \text{ kJ}}$$

> ### 🔍 Hint
> From $E_h = cm\Delta T$
> $$c = \frac{E_h}{m\Delta T}$$
> $$m = \frac{E_h}{c\Delta T}$$
> $$\Delta T = \frac{E_h}{cm}$$

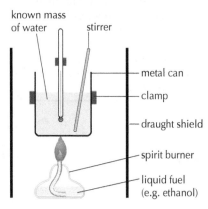

known mass of water   stirrer
— metal can
— clamp
— draught shield
— spirit burner
— liquid fuel (e.g. ethanol)

Figure 2.6.11: *Simple laboratory calorimeter*

The energy produced by burning different alcohols can be compared using a simple calorimeter system that can be built in the laboratory.

To ensure that as much of the energy released by burning the fuel is transferred to the water a metal can, preferably made of copper, is used. Draught shields are placed around the burner to minimise heat loss due to draughts and air convection.

Procedure for carrying out each experiment:

- The metal can is clamped at a suitable height. There needs to be a gap of about 2–5 cm between the top of the burner wick and the base of the can.

- 100 cm³ of water is poured into the can from a measuring cylinder.

- The initial temperature of the water is recorded.

- The spirit burner is placed on the heat-resistant mat under the can, the cap removed and the burner lit.

- The water is stirred gently. When the temperature has risen by about 20 °C the burner is capped to extinguish the flame.

- The final temperature of the water is recorded. The thermometer reading may continue to rise for a few seconds after the flame has been extinguished.

GO! Activity 2.6.7

1. The table below shows measurements a student recorded when calculating energy produced by burning methanol, ethanol and propan-1-ol.

	Methanol	Ethanol	Propan-1-ol
Initial temperature of water /°C	23·2	22·8	23·1
Final temperature of water /°C	45·6	46·3	47·2
ΔT/°C	22·4	23·5	24·1
$E_h = cm\Delta T$ (Energy transferred to water)	9·36 kJ	(a)	(b)

Calculate the values for (a) and (b).

2. Data obtained from an experiment to find the specific heat capacity of a liquid is shown in the table below.

Measurement	Result
Enthalpy change (kJ)	5.1
Mass of liquid heated (g)	100
Initial temperature of liquid (°C)	21.1
Final temperature of liquid (°C)	42.3

Calculate the specific heat capacity (c) of the liquid, in kJ kg^{-1} °C^{-1}.

3. Ethanol was burned to produce energy to raise the temperature of a sample of water. The heat energy, E_h, required to raise the temperature of water by 21 °C was 2.55 kJ.

Calculate the mass of water which was heated.

4. A student heated 200 g of water using an ethanol burner. The student calculated that 10.2 kJ of energy had been transferred to the water during the experiment. Calculate the temperature rise that the student would have recorded.

Word bank

- **Combustion**

The reaction of a fuel with oxygen producing energy (exothermic).

- **Exothermic**

A reaction in which energy is given out.

Combustion reactions give out energy to the surroundings. Any reaction which gives out energy is described as **exothermic**.

In order for some reactions to take place energy needs to be taken in from the surroundings. The surroundings become cooler as they lose energy. Reactions which take in energy from the surroundings are described as **endothermic**.

Example 2.6.3

Word bank

- **Endothermic**

Energy is taken in from the surroundings during a chemical reaction.

When 1 g of solid ammonium chloride was added to 20 cm³ of water, the temperature of the water dropped from 21.4 °C to 18·5 °C. Calculate the energy lost by the water.

Worked answer:

$c = 4.18$ kJ kg⁻¹ °C⁻¹

$m = 0.02$ kg (20 cm³ = 20 g = 0·02 kg) – remember the mass must be changed to kilograms

$\Delta T = 2.9$ °C

Energy lost by water = $E_h = cm\Delta T$
$= 0.02 \times 4.18 \times 2.9$
$E_h = 0.24$ kJ

Carboxylic acids

Carboxylic acids are a homologous series that contains the carboxyl group: $-COOH$

carboxyl group

Word bank

- **Carboxylic acids**

Acids containing the carboxyl functional group.

Citric acid is a carboxylic acid found in citrus fruits such as lemon. Lactic acid, found in yoghurts, is made by fermenting lactose, a sugar found in milk. Lactic acid is also produced by the body. This causes the burn that people feel during strenuous exercise.

Vinegar is probably the best-known substance to contain a carboxylic acid. Vinegar is a solution of ethanoic acid in water.

ethanoic acid

◄◄◄ SPOTLIGHT ON VINEGAR

Go into any supermarket and you will find a variety of different types of vinegar: malt vinegar, red wine vinegar, white wine vinegar and balsamic vinegar, to name but a few. In some countries fruit vinegars are made by sweetening the vinegar with fruit or fruit juice. In the Middle East vinegars made from dates are popular and coconut vinegar is used in the Far East.

The word vinegar comes from the French words 'vin' meaning wine and 'aigre' which means sour. Vinaigre therefore means sour wine.

It didn't take long for us to discover vinegar after we had discovered wine. Wine will naturally turn to vinegar if left to stand, as bacteria oxidise the ethanol in the wine to ethanoic acid.

Very quickly vinegar was found to be versatile and extremely useful. From earliest recorded times vinegar has been used. In ancient Babylonia it was used as a cleaning agent and as a food preservative. Legend has it that the Egyptian queen, Cleopatra, demonstrated vinegar's ability as a solvent when at the end of meal she dissolved a pearl in vinegar and drank the solution to prove she could consume a fortune in a single meal. Hannibal, the Carthaginian General, marched an army with elephants over the Alps in order to invade Rome. The army removed rocks blocking their route by heating the rocks and pouring vinegar over them. The rocks became crumbly and could be chipped away allowing the army to pass. Helen of Troy is reputed to have bathed in vinegar to relax.

In more recent times vinegar was used by soldiers in the American Civil War to clean wounds.

Today vinegar still has many varied uses. These range from cooking to cleaning, from soothing insect bites to dealing with unpleasant smells.

> ### 🔍 Hint
>
> A simple internet search will allow you to find out more about the history and varied uses of vinegar.

Figure 2.6.12: *Bottles of assorted aromatic vinegars with some of the herbs and flowers used to flavour them*

Using carboxylic acids

Carboxylic acids are used as preservatives. Pickling is commonly used to preserve food such as onions and beetroot. The low pH of vinegar stops the growth of harmful bacteria and fungi. Vinegar is also used in household cleaning products as it is non-toxic. Carboxcylic acids are also used in the manufacture of soaps and some medicines.

Carboxylic acids – a homologous series

Ethanoic acid, the acid in vinegar, is the second member of a homologous series of carboxylic acids based on the alkanes.

The carboxyl group, formula –COOH, occurs at the end of a chain of carbons.

The simplest carboxylic acid is methanoic acid. It was known as formic acid and derived its name from 'formica', the Latin for ant. Ant and bee stings contain this acid.

methanoic acid

Carboxylic acids based on straight-chain alkanes are named by dropping the -e from the alkane name and adding -oic acid to the stem, e.g. the carboxylic acid with 5 carbons is pentanoic acid.

pentane   pentanoic acid

The general formula for the carboxylic acids based on alkanes can be written as $C_nH_{2n+1}COOH$, where n = 0, 1, 2, 3 etc.

(Be careful! In this case 'n' is not related to the name since there is also a carbon in the carboxyl group.)

Example 2.6.4

Draw the full and shortened structural formulae for the carboxylic acid with molecular formula C_3H_7COOH. Name the acid.

full structural formula:

shortened structural formula: $CH_3CH_2CH_2COOH$

name: butanoic acid (the acid has 4 carbons)

Like the alcohols, straight-chain carboxylic acids show a gradation in some physical properties.

Physical properties of carboxylic acids

Table 2.6.7 shows the boiling points and solubilities of the first eight carboxylic acids.

Table 2.6.7: *Boiling points and solubilities of the carboxylic acids*

Carboxylic acid	Shortened structural formula	Boiling point/°C	Solubility in water/gl⁻¹
methanoic acid	HCOOH	101	miscible
ethanoic acid	CH_3COOH	118	miscible
propanoic acid	CH_3CH_2COOH	140	miscible
butanoic acid	$CH_3(CH_2)_2COOH$	163	60
pentanoic acid	$CH_3(CH_2)_3COOH$	186	24
hexanoic acid	$CH_3(CH_2)_4COOH$	202	10
heptanoic acid	$CH_3(CH_2)_5COOH$	223	3
octanoic acid	$CH_3(CH_2)_6COOH$	238	1

As the molecules get bigger, the boiling points increase. This is because it takes more energy to separate the molecules. As the molecules get bigger, their solubility decreases. This is because bigger molecules can't interact with water molecules as well as smaller molecules can.

Chemical properties of carboxylic acids

Solutions of carboxylic acids have a pH less than 7 and, like other acids, can react with metals and bases (metal oxides, hydroxides and carbonates) forming salts.

Example 2.6.5

methanoic acid + magnesium → magnesium methanoate + hydrogen

$$HCOOH + Mg → (HCOO)_2Mg + H_2$$

Example 2.6.6

ethanoic acid + sodium oxide → sodium ethanoate + water

$$2CH_3COOH + Na_2O → 2CH_3COONa + H_2O$$

Example 2.6.7

propanoic acid + potassium hydroxide → potassium ethanoate + water

$$C_2H_5COOH + KOH → C_2H_5COOK + H_2O$$

Example 2.6.8

butanoic acid + calcium carbonate → calcium butanoate + water + carbon dioxide

$$2C_3H_7COOH + CaCO_3 → (C_3H_7COO)_2Ca + H_2O + CO_2$$

The reaction of carboxylic acids with bases is an example of neutralisation.

A pickled egg can be made by placing a hard-boiled egg still in its shell in vinegar. The vinegar reacts with the calcium carbonate of the shell leaving behind the egg.

Figure 2.6.13: *Pickling can be used to preserve food such as eggs*

Figure 2.6.14: *The build-up of limescale in a pipe will impair flow and decrease heat transfer through the wall of the pipe*

In hard-water areas limescale can build up on the heating elements in boilers and kettles. One of the ways of dealing with this is to use vinegar. Limescale is mainly calcium carbonate. The ethanoic acid in the vinegar reacts with the limescale to give calcium ethanoate, carbon dioxide and water. The calcium ethanoate is soluble and can be rinsed away.

$$\text{ethanoic acid} + \text{calcium carbonate} \rightarrow \text{calcium ethanoate} + \text{water} + \text{carbon dioxide}$$

$$2CH_3COOH + CaCO_3 \rightarrow (CH_3COO)_2Ca + H_2O + CO_2$$

GO! Activity 2.6.8

1. Name the following compounds.

   (a)

   ```
        H   H   H       O
        |   |   |      //
    H—C—C—C—C
        |   |   |      \
        H   H   H       O—H
   ```

   (b)

   ```
        H   H   H   H   H   H       O
        |   |   |   |   |   |      //
    H—C—C—C—C—C—C—C
        |   |   |   |   |   |      \
        H   H   H   H   H   H       O—H
   ```

2. (a) Draw the structural formulae for the following carboxylic acids.

   Pentanoic acid

   Octanoic acid

   (b) Write the molecular formula for each of the compounds in part (a).

3. Complete the following word equations and write formula equations for each.

(a) ethanoic acid + zinc →

(b) butanoic acid + potassium oxide →

(c) propanoic acid + lithium hydroxide →

(d) methanoic acid + copper(II) carbonate →

4. Give three uses for carboxylic acids.

Learning checklist

In this chapter you have learned:

- Alcohols are a group of compounds containing the hydroxyl (-OH) functional group.

- To write molecular formulae and draw structural formulae for straight-chain alcohols given their name.

- To name straight-chain alcohols given their shortened or full structural formulae.

- Alcohols show a gradation in their physical properties.

- Alcohols are very good solvents.

- Alcohols burn with a very clean flame and can be used as fuels.

- Combustion, burning of fuels, is an exothermic process.

- Endothermic reactions take in energy from their surroundings.

- Fuels can be compared by measuring the energy given off when they are burned.

- When a burning fuel is used to heat water the energy transferred is calculated using the formula $E_h = cm\Delta T$.

- Carboxylic acids are a homologous series of compounds containing the carboxyl (COOH) functional group.

- To write molecular formulae and draw structural formulae for straight-chain carboxylic acids given their name.

- To name straight-chain carboxylic acids given their shortened or full structural formulae.

- Vinegar is a solution of ethanoic acid in water.

- Vinegar can be used as a preservative.

- Many household cleaners contain vinegar.

- Solutions of carboxylic acids have pH less than 7.

- Carboxylic acids react with metals and bases to form salts.

- How to name the salts formed from carboxylic acids.

AREA 3

Chemistry
in society

7 Metals

You should already know

- The chemical and physical properties of elements are related to their position in the Periodic Table.
- Metals lie to the left-hand side of the Periodic Table.
- The reactions of metals with oxygen, water and dilute acid can be used to deduce a reactivity series for the metals.
- The method used to extract a metal from its ores depends on the reactivity of the metal.
- Corrosion of metals takes place when the surface of the metal reacts to form compounds.
- Metals can be protected from corrosion by physical and chemical means.
- An electric current flows when different metals are connected and placed in an electrolyte.
- Comparing the voltage readings between pairs of metals can be used to deduce an electrochemical series.
- The electrochemical series can be used to predict the voltage and current direction in an electrochemical cell.
- Formulae for ionic compounds can be represented showing ion charges.

Learning intentions

In this chapter you will learn about:
- Extraction of metals
  - > heating
  - > smelting
  - > using electricity.
- Metallic bonding.
- Reactions of metals and redox reactions.
- Electrochemical cells.

STEP BACK IN TIME: METALS THROUGH THE AGES

The ability of humans to process metals and create metal alloys has been fundamental to the advancement of civilisation.

Until the beginning of the thirteenth century only seven metals were known. These metals, known as the Metals of Antiquity, were gold, silver, copper, lead, tin, iron and mercury. These were metals that were either found uncombined or were easily extracted from their **ores** *by a* **smelting** *process.*

Gold and copper were the first metals to be known and used. Gold was found as small nuggets in streams and simply processed by hammering the pieces together into jewellery.

Humans then learned to smelt metals from their ores. Ores are rocks that contain minerals from which the metals can be obtained. It is thought that the process of smelting may have been developed after small amounts of metals were found in potters' kilns.

It was also found that copper could be made harder and more durable by smelting a mixture of different ores. Some of the minerals in the ores contained tin. The new metal, a mixture of copper with some tin, was called bronze, the first man-made alloy.

> 📖 **Word bank**
>
> • **Ores**
> Rocks from which metals can be obtained.

> 📖 **Word bank**
>
> • **Smelting**
> Reducing a metal ore by heating with carbon.

Figure 3.7.1: *Ancient Roman bronze coins*

Figure 3.7.2: *A bronze ornament dating from the fourth century BCE*

It wasn't until the eighteenth century that scientists found new metals that could be obtained by smelting. The discovery of electricity then allowed the more reactive metals to be obtained. Eighty-six elemental metals are now known.

> ## 📖 Word bank
>
> ● **Heavy metals**
> Elements such as cadmium and mercury, which are toxic to living things.

The discovery of metals and the method used to extract them are related to their reactivity. The metals used in antiquity were metals that were lower in reactivity.

The reactive metals sodium and potassium were both discovered in 1807 by the famous British scientist Sir Humphry Davy who was investigating the electrolysis of solutions.

And now in the twenty-first century – Going for Gold!

The gold medals for the 2012 Olympics in London only contained about 6 grams of gold. The gold was used to plate the medals.

The last Olympic gold medals actually made of solid gold were awarded in 1912.

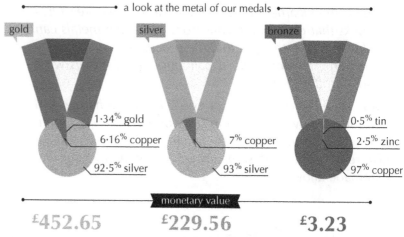

a look at the metal of our medals

gold
1·34% gold
6·16% copper
92·5% silver

silver
7% copper
93% silver

bronze
0·5% tin
2·5% zinc
97% copper

monetary value
£452.65   £229.56   £3.23

Figure 3.7.4: *The actual cost of the medals*

Figure 3.7.3: *2012 Olympic medals*

The organisers of the 2012 Olympics were offered 12 g of Scottish gold panned from a Highland burn but turned down the offer. The medals for the 2014 Commonwealth Games in Glasgow were of a similar composition to the London Olympic medals.

Extraction of metals

Some metals like gold can be found uncombined in the Earth's crust. Most metals, however, exist in the Earth's crust as compounds. Rocks from which metals can be obtained are known as ores.

Figure 3.7.5: *Hematite is an important ore of iron. It is also an important pigment and was used in ancient times to create cave paintings*

Figure 3.7.6: *A specimen containing the copper ore azurite. This compound was also used as pigment*

Example 3.7.1

Chemical analysis of an ore provides information that allows the mass of metal that can be obtained from the ore to be worked out. Analysis of hematite shows it is iron(III) oxide (Fe_2O_3). The percentage composition by mass of iron present in hematite can be worked out from the formula:

$$\% \text{ composition by mass of Fe} = \frac{\text{mass of Fe in formula}}{\text{formula mass of } Fe_2O_3} \times 100$$

$$= \frac{(2 \times 56)}{(2 \times 56) + (3 \times 16)} \times 100$$

$$= \frac{112}{160} \times 100$$

% composition by mass of Fe = 70%

GO! Activity 3.7.1

Copper can be obtained from the ore malachite. The formula for malachite is $Cu_2CO_3(OH)_2$.

Calculate the percentage by mass of copper in malachite ore.

The method of extraction of a metal from its ore depends on how reactive a metal is.

K Na Li Ca Mg Zn Fe Sn Pb Cu Ag Au Pt

Most reactive ————→ Least reactive

The order of how reactive a metal is is known as the reactivity series (see page 150).

Heating

Some metals low in reactivity can be obtained simply by heating their ore.

If silver oxide is heated silver metal is produced.

$$2Ag_2O(s) \rightarrow 4Ag(s) + O_2(g)$$

Ionic equation: $2(Ag^+)_2O^{2-}(s) \rightarrow 4Ag(s) + O_2(g)$

Silver oxide is an ionic compound.

The metal ore is said to have been reduced to the metal.

During the reaction the positive silver ions change to atoms by gaining an electron. Gain of electrons is called **reduction**.

This can be shown in an ion-electron equation. Ion-electron equations show the electrons gained or lost by an atom or ion. The SQA data book gives a list of commonly used ion-electron equations.

$$Ag^+(s) + e^- \rightarrow Ag(s) \qquad \textit{reduction}$$

Smelting

Smelting is a process for extracting a metal from its ore by mixing with a source of carbon and heating. The best-known example of smelting is that of iron in a blast furnace.

> ### Word bank
> • **Reduction**
> Gain of electrons.

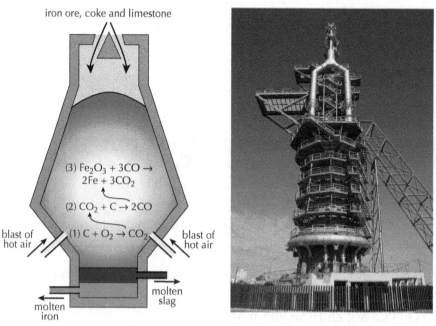

iron ore, coke and limestone

(3) $Fe_2O_3 + 3CO \rightarrow 2Fe + 3CO_2$

(2) $CO_2 + C \rightarrow 2CO$

(1) $C + O_2 \rightarrow CO_2$

blast of hot air

blast of hot air

molten iron

molten slag

Figure 3.7.7: *Inside and outside a blast furnace*

Iron ore, coke (a source of carbon) and limestone are loaded in from the top of the furnace and hot air is blown in from the base.

Coke reacts with the oxygen of the air being blown in to give carbon dioxide.

(1) $C(s) + O_2(g) \rightarrow CO_2(g)$

This reacts with more coke producing carbon monoxide.

(2) $C(s) + CO_2(g) \rightarrow 2CO(g)$

The iron ore reacts with the carbon monoxide and is reduced to iron.

(3) $Fe_2O_3(s) + 3CO(g) \rightarrow 2Fe\ (\ell) + 3CO_2(g)$

An ion-electron equation can be written for the production of iron. Iron ions in the iron oxide are reduced to iron atoms when they gain electrons.

$$Fe^{3+}(\ell) + 3e^- \rightarrow Fe(\ell)$$

The iron forms at the bottom of the furnace and can be run off.

GO! Activity 3.7.2

Heating copper(II) oxide with carbon produces copper and carbon dioxide.

1. Write a balanced equation for the reaction.
2. Write an ion-electron equation for the reduction of copper ions to copper atoms.

📖 Word bank

• **Electrolysis**

Breaking down an ionic compound by passing a d.c. current through it.

Using electricity

The most reactive metals are extracted using electricity.

In the lab the use of electricity to extract a metal can be demonstrated by the **electrolysis** of molten lead bromide. Electrolysis is the breaking down of a compound using electricity. A **d.c. supply** has to be used so that the products can be identified.

📖 Word bank

• **Direct current (d.c.)**

Gives a positive and negative electrode so products of electrolysis can be identified.

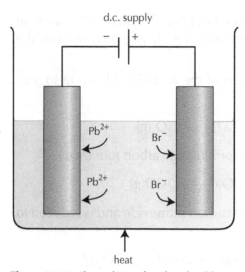

Figure 3.7.8: *Electrolysis of molten lead bromide*

During the electrolysis reddish brown bromine vapour is seen around the positive electrode.

Lead is seen at the negative electrode.

$$Pb^{2+}(\ell) + 2e^- \rightarrow Pb(\ell) \qquad reduction$$

Aluminium is obtained by electrolysing molten aluminium oxide. The positive aluminium ions are attracted to the negative electrode, where they are reduced to atoms.

$$Al^{3+}(\ell) + 3e^- \rightarrow Al(\ell) \qquad reduction$$

In each example, the negative ions lose electrons at the positive electrode.

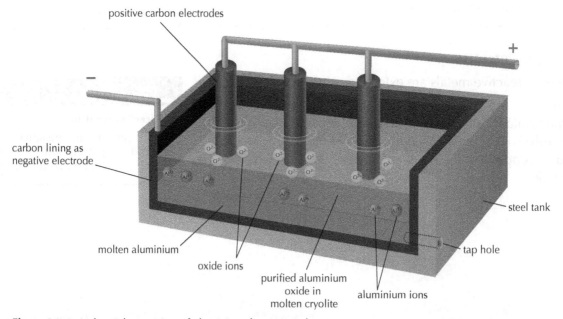

Figure 3.7.9: *Industrial extraction of aluminium from its molten ore*

Aluminium has been termed 'The Wonder Metal'. This is because of the extremely diverse properties and uses of the metal and its alloys.

Aluminium has several key properties. These include:

- *Low density – aluminium is only one-third the weight of steel.*

- *Corrosion resistance – a natural coating of aluminium oxide makes the metal aluminium highly resistant to most forms of corrosion.*

- *Excellent electrical conductivity – aluminium is used in high voltage electricity transmission lines.*

- *Non-toxicity – aluminium is used widely in food packaging industries.*

Commercial manufacture of aluminium began in 1886. The main ore of aluminium is bauxite, which has a reddish colour due to the presence of iron oxide. It is refined to separate it from impurities, giving the pure mineral alumina, which can then be electrolysed to give aluminium.

World aluminium production has grown steadily and continues to increase.

Figure 3.7.10: *An aluminium smelter*

There are estimated to be 55–75 billion tonnes of bauxite available – meaning that the Earth is not going to run out of aluminium anytime soon.

Table 3.7.1 shows how aluminium production in China grew from 2001 to 2010.

Table 3.7.1: *Table of aluminium production for China 2001–2010*

Year	Aluminium (millions of tonnes)
2001	3·4
2002	4·3
2003	5·5
2004	6·7
2005	7·8
2006	9·3
2007	12·6
2008	13·1
2009	13·0
2010	16·1

GO! Activity 3.7.3

1. Use the data in Table 3.7.1 to draw a bar chart of aluminium production for China from 2001 to 2010.

2. Use the data to predict how much aluminium China produced in 2015.

3. Aluminium is produced from alumina (aluminium oxide, Al_2O_3).

$$2Al_2O_3(s) \rightarrow 4Al(\ell) + 3O_2(g)$$

If all the alumina was converted to aluminium, what mass of alumina would have been used, on average, each month in 2010?

🔍 Hint

An internet search will allow you to find out the latest developments for using aluminium alloys in aircraft manufacture and other uses.

One of the main uses of aluminium and aluminium alloys is in aircraft construction. There are over 27 000 commercial aircraft flying in the world as well as many thousands of light aircraft and helicopters. Demand for passenger aircraft alone is set to double from 15 000 to more than 31 500 by 2030.

Aluminium is the main aircraft material and comprises about 80% of an aircraft's unladen weight. Because the metal resists corrosion, some airlines don't paint their planes, saving several hundred of kilograms in weight.

Aluminium alloys are categorised by number. Aluminium 7075 is an alloy containing zinc, magnesium and copper. It is a high-strength alloy used by aircraft manufacturers to strengthen aluminium aircraft structures. The natural oxide coating on aluminium can be thickened by a process known as anodising. The oxide coating can be dyed. Aluminium 7075 anodises beautifully. It also has very good machinability, giving an excellent finish.

Aluminium–lithium alloys are now being developed for use by aircraft manufacturers in the next generation of aircraft.

Metallic bonding

Metals have certain properties in common.

- They are good conductors of heat and electricity.
- They tend to be shiny (have metallic lustre).
- They can be shaped (are malleable).
- They can be drawn into wires (are ductile).

These properties of metals can be explained by the bonding within the metals.

In metals the outer electrons of the atoms can move easily from atom to atom. The electrons within the structure are said to be **delocalised** (this simply means they are not confined to a particular region of space). Metals structures can be described as 'positive ions in a sea of electrons'.

> **📖 Word bank**
>
> • **Delocalised electrons**
> Electrons which can move easily from one atom to another.

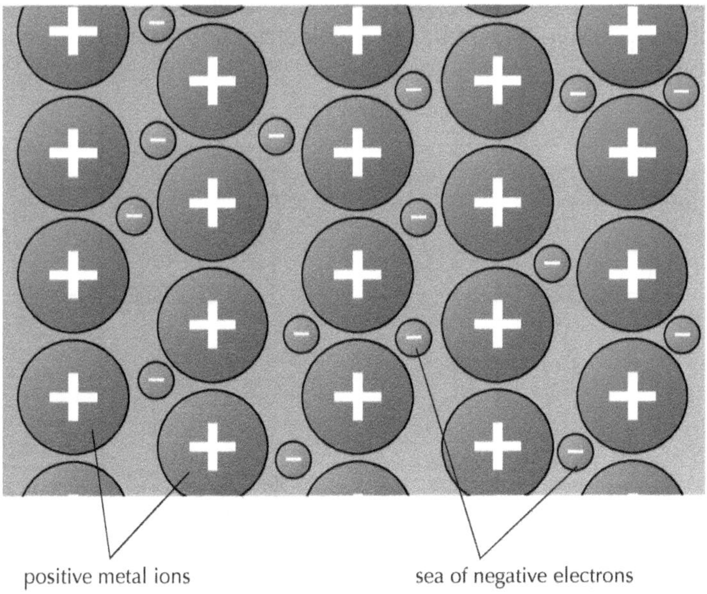

positive metal ions   sea of negative electrons

Figure 3.7.11: *Metals exist as positive ions in a sea of delocalised electrons*

The structure is held together by metallic bonds. These are the attractive forces of the metal nuclei for the delocalised electrons moving between them. The direction of the bonds is not fixed because the electrons are moving, unlike in covalent compounds where the electrons are localised in bonds. This means that the atoms will be able to move in relation to each other and explains why the metals can be rolled into thin sheets or drawn into wires.

Figure 3.7.12: *Red-hot roll of steel in a steel mill*

Metals are good electrical conductors because they contain delocalised electrons. If a voltage is placed across a metal the delocalised electrons will move from the negative terminal towards the positive terminal, i.e. an electrical current will flow through the metal.

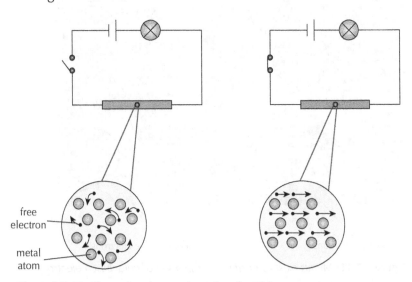

free
electron

metal
atom

Figure 3.7.13: *Illustrating the conductivity of metal*

Reactions of metals and redox reactions

The order of reactivity of metals can be established by the reaction of the metals with oxygen, water and acids.

Metals and oxygen

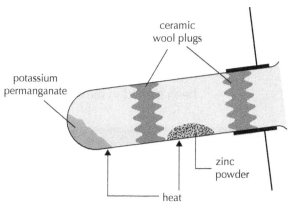

Figure 3.7.14: *Reacting zinc with oxygen*

When zinc reacts with oxygen, zinc oxide is produced.

Balanced equation: $2Zn(s) + O_2(g) \rightarrow 2ZnO(s)$

Ionic equation: $2Zn(s) + O_2(g) \rightarrow 2Zn^{2+}O^{2-}(s)$

The zinc atoms are changed to zinc ions by losing two electrons.

$Zn(s) \rightarrow Zn^{2+}(s) + 2e^-$ This change is known as **oxidation**. Oxidation is the loss of electrons by a molecule, atom or ion.

> **Word bank**
>
> • **Oxidation**
> Loss of electrons.

The electrons lost by the zinc are gained by the oxygen changing oxygen molecules to oxide ions.

$O_2(g) + 4e^- \rightarrow 2O^{2-}(s)$ This change is known as reduction. Reduction is the gain of electrons by a molecule, atom or ion.

> **Word bank**
>
> • **Redox equation**
> Reduction and oxidation ion – electron equations added together.

The two ion-electron equations for the oxidation and the reduction parts of the reaction can be combined to give a **redox equation** for the whole reaction by balancing the electrons lost and gained.

$Zn(s) \rightarrow Zn^{2+}(s) + 2e^-$ This equation for the oxidation is multiplied by 2 to give $4e^-$, the same number gained by the oxygen.

> **Word bank**
>
> • **Redox reaction**
> Reaction in which loss and gain of electrons takes place.

$$2Zn(s) \rightarrow 2Zn^{2+}(s) \ + 4e^-$$
$$O_2(g) \ + 4e^- \rightarrow 2O^{2-}(s)$$

_____

$2Zn(s) + O_2(g) \rightarrow 2Zn^{2+}O^{2-}(s)$ Adding the oxidation half-equation and the reduction half-equation gives the redox equation for the reaction. Notice that no electrons are shown in the redox equation because those lost by the zinc have been gained in the oxygen.

Activity 3.7.4

Magnesium metal powder is used in fireworks and flares to give a brilliant white light.

The magnesium reacts with oxygen in the air to produce magnesium oxide.

1. Write ion-electron equations for:
   (a) the oxidation of magnesium
   (b) the reduction of oxygen.
2. Combine the oxidation and reduction equations to give the redox equation for the reaction.

Figure 3.7.15: *Lithium reacting with water with universal indicator*

Metals reacting with water and acid

Metals and water

Alkali metals (Group 1 in the Periodic Table) are so called because they react with water to give alkaline solutions.

When lithium reacts with water the lithium atoms are oxidised. Water is reduced to hydrogen gas and hydroxide ions.

$$Li(s) \quad \rightarrow \quad Li^+(aq) + e^- \qquad \textit{oxidation}$$
$$2H_2O(\ell) + 2e^- \quad \rightarrow \quad H_2(g) \ + 2OH^-(aq) \quad \textit{reduction}$$

The equations can be combined by balancing the electrons lost and gained to give the overall redox equation for the reaction.

$$2Li(s) \ + \ 2H_2O(\ell) \rightarrow H_2(g) \ + \ 2Li^+(aq) + 2OH^-(aq) \quad \textit{redox}$$

GO! Activity 3.7.5: Paired activity

Sodium is more reactive than lithium. It also reacts with water producing an alkaline solution.

Working with a partner:

1. Write ion-electron equations for:
   (a) the oxidation of the sodium atoms
   (b) the reduction of the water.
2. Combine the equations for oxidation and reduction to give a balanced redox equation for the reaction.

⚠ Remember!

Ion-electron equations can be found in the SQA data booklet.

⚠ Remember!

Only the atoms/molecules/ions reacting are shown in reduction, oxidation and redox equations.

Metals and acid

When zinc reacts with dilute hydrochloric acid, zinc atoms are oxidised and hydrogen ions in the acid are reduced to hydrogen gas, which is given off.

$$Zn(s) \quad \rightarrow \quad Zn^{2+}(aq) + 2e^- \quad \textit{oxidation}$$
$$2H^+(aq) \ + \ 2e^- \quad \rightarrow \quad H_2(g) \qquad \textit{reduction}$$

The ion-electron equations can be combined to give the redox equation for the reaction.

$$Zn(s) \ + \ 2H^+(aq) \quad \rightarrow \quad Zn^{2+}(aq) \ + H_2(g) \quad \textit{redox}$$

Metals which react with acids can be used to make soluble salts. Excess metal is added to the acid, the mixture is filtered and the filtrate evaporated to dryness.

Figure 3.7.16: *Zinc reacting with dilute hydrochloric acid*

GO! Activity 3.7.6

Iron reacts with dilute sulfuric acid to give hydrogen gas.

1. Write ion-electron equations for:
   (a) the oxidation of the iron atoms
   (b) the reduction of hydrogen ions in the acid.
2. Combine the equations for oxidation and reduction to give a balanced redox equation for the reaction.

The reactivity series

The results of how reactive metals are with oxygen, water and acid enables us to put metals in order of how reactive they are. This is known as the reactivity series.

K Na Li Ca Mg Zn Fe Sn Pb Cu Ag Au Pt

Most reactive ─────→ Least reactive

Electrochemical cells

🚲 STEP BACK IN TIME

In 1780 Luigi Galvani demonstrated that when two different metals were connected and touched against the nerve in a frog's leg, the leg would twitch. This was the first step on the road to understanding that connecting different metals would produce electricity. In 1800 Alessandro Volta invented the voltaic pile, the world's first battery. This in turn led to Humphry Davy using the pile to discover new metal elements.

A simple voltaic pile can be made using pennies, zinc washers and pieces of card or filter paper soaked in salt solution. The pile can then be used to light an LED.

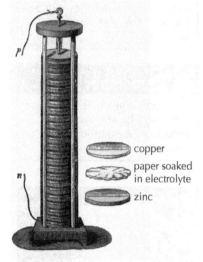

copper

paper soaked in electrolyte

zinc

Figure 3.7.17: *An early voltaic pile*

Cells involving metals

A cell can be made by dipping zinc and copper rods in solutions containing their own ions and joining the solutions using an **ion bridge**. An ion bridge can be made by soaking a filter paper in a salt solution. A salt solution is an example of an ionic solution which conducts electricity and is know as an **electrolyte**. The ion bridge allows ions to travel through the solutions carrying the current.

Electrons flow through the wire from the zinc electrode to the copper electrode. Zinc is higher in the electrochemical series (see page 153) than copper, therefore will lose electrons more easily.

📖 Word bank

• **Ion bridge**

An electrolyte which allows ions to flow between the two solutions in an electrochemical cell.

📖 Word bank

• **Electrolyte**

An ionic solution which conducts electricity.

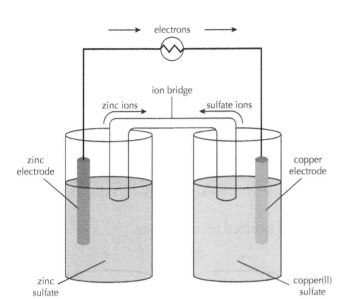

Figure 3.7.18: *A zinc-copper cell*

Each beaker is known as a half cell and the reaction taking place in the beaker is described by an **ion-electron equation**.

At the zinc electrode zinc atoms are oxidised to zinc ions producing electrons.

$$Zn(s) \rightarrow Zn^{2+}(aq) + 2e^- \quad oxidation$$

The electrons flow through the wires to the copper where copper ions from the solution gain electrons and change to copper atoms. This is reduction.

$$Cu^{2+}(aq) + 2e^- \rightarrow Cu(s) \quad\quad reduction$$

The redox equation for the overall cell reaction is formed by combining the two half-equations.

$$Zn(s) + Cu^{2+}(aq) \rightarrow Zn^{2+}(aq) + Cu(s) \quad redox$$

This equation doesn't show the sulfate ions which are also present. These are known as spectator ions since they are present but are not involved in the reaction.

The overall ionic equation showing spectator ions:

$$Zn(s) + Cu^{2+}(aq) + SO_4^{2-}(aq) \rightarrow Zn^{2+}(aq) + SO_4^{2-}(aq) + Cu(s)$$

📖 **Word bank**

• **Ion-electron equation**
Equations which show the electrons gained or lost by an atom or ion. The SQA data book gives a list of commonly used ion-electron equations.

🛈 Activity 3.7.7

An electrochemical cell can be made using magnesium and tin as electrodes, and solutions of their ions.

(a) Draw and label a diagram of the cell.

(b) On the cell indicate the direction of electron flow.

Cells involving non-metals

Since all redox reactions involve transfer of electrons any redox reaction has the potential to produce an electric current.

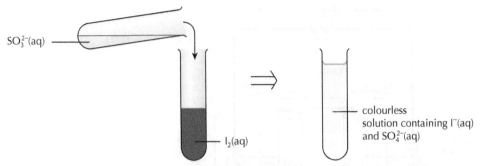

Figure 3.7.19: *A redox reaction takes place when sulfite ions are added to iodine solution*

Iodine molecules are reduced to iodide ions.

$$I_2(aq) \quad + \quad 2e^- \quad \rightarrow \quad 2I^-(aq)$$

reddish brown   *colourless*

This reaction will give a voltage if it is carried out in a cell. The solutions are placed in separate beakers and the electrodes in the cell are made of **graphite** (carbon). The graphite conducts electricity but is inert, i.e. doesn't react with the solutions.

The solutions in the beakers are joined by an ion bridge which allows ions to move between the solutions, completing the circuit.

> ### 📖 Word bank
>
> • **Graphite**
> A form of carbon often used as electrodes in an electrochemical cell because it conducts electricity and will not react with the electrolytes.

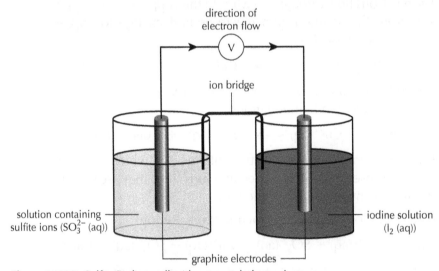

Figure 3.7.20: *Sulfite/iodine cell without metal electrodes*

The sulfite ions are oxidised to sulfate ions.

The ion-electron half equation for the oxidation is:

$$SO_3{}^{2-}(aq) + H_2O(\ell) \rightarrow SO_4{}^{2-}(aq) + 2H^+(aq) + 2e^- \quad \textit{oxidation}$$

The electrons flow through the circuit and the iodine molecules are reduced at the other electrode.

$$I_2(aq) \quad + \quad 2e^- \quad \rightarrow \quad 2I^-(aq) \qquad\qquad\qquad \textit{reduction}$$

The redox equation for the reaction is formed by combining the oxidation and reduction ion-electron half equations:

$$SO_3^{2-}(aq) + H_2O(\ell) + I_2(aq) \rightarrow SO_4^{2-}(aq) + 2I^-(aq) + 2H^+(aq)$$

The electrochemical series

When different metals are used to make an electrochemical cell, different voltages are produced. Figure 3.7.21 shows how a simple cell can be used to obtain voltages.

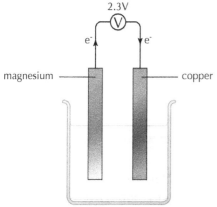

Figure 3.7.21: *Chemical cell showing direction of electron (e⁻) flow from magnesium to copper, through the wire*

Table 3.7.2 shows the typical voltage obtained when different metals are connected to copper.

The voltages obtained can be used to put the metals in order of how easily they give away electrons (form ions). This is known as an **electrochemical series** and is shown in Table 3.7.3.

The further apart metals are in the electrochemical series, the higher the voltage produced when they are connected in a cell. Electrons flow through the wires from the metal higher in the series to the lower metal.

Table 3.7.2: *Typical voltages obtained when different metals are connected to copper.*

Metal connected to copper	Voltage (V)
magnesium	2.3
zinc	1.4
iron	0.9
lead	0.5

Table 3.7.3: *The electrochemical series*

Metal
lithium
potassium
calcium
sodium
magnesium
aluminium
zinc
iron
nickel
tin
lead
copper
silver
gold

> ### ⓖ Activity 3.7.8
>
> A cell can be made using graphite electrodes in iron(II) sulfate ($Fe^{2+}SO_4^{2-}$) solution and acidified potassium permanganate solution ($K^+MnO_4^-$). A rolled up filter paper soaked in potassium nitrate solution makes a suitable ion bridge.
>
> A meter will show that electrons flow through the wires from the beaker containing the iron(II) ions to the beaker containing the permanganate ions.

> ### 📖 Word bank
>
> • **Electrochemical series**
> Metals listed in order of how easily they give away electrons.

Hint

The electrochemical series found in data booklets includes ions which gain or lose electrons as well as metals.

The permanganate ions are reduced:

$$MnO_4^-(aq) + 8H^+(aq) + 6e^- \rightarrow Mn^{2+}(aq) + 4H_2O(\ell)$$

1. Draw and label a diagram of the cell.
2. Write the ion-electron equation for the oxidation of $Fe^{2+}(aq)$. (You may wish to use the SQA data booklet.)
3. Combine the oxidation and reduction ion-electron half equations to give the overall redox equation for the cell.

◀€ SPOTLIGHT ON TECHNOLOGY

Fuel cells

A fuel cell is an electrochemical cell that produces electricity by combining a fuel such as hydrogen with oxygen without burning it.

The use of fuel cells is increasing. They represent a clean technology which will help industries and governments meet carbon dioxide emission reduction targets.

Figure 3.7.22: *Promoting Unst Renewable Energy (PURE) project linked wind turbines to fuel cells*

A project in Unst in Shetland linked wind turbines to fuel cells. The electricity from the wind turbines was used to electrolyse water to give hydrogen and oxygen which could be stored. The gases were then used in fuel cells to produce electricity when required.

One type of hydrogen fuel cell consists of two electrodes with a special membrane (proton exchange membrane) which allows hydrogen ions (protons) to move through and acts as the electrolyte in the cell.

At one electrode hydrogen reacts with a catalyst, creating positively charged hydrogen ions and electrons. The ions then pass through the proton exchange membrane while the electrons travel through the external circuit creating a current. At the other electrode oxygen reacts with the hydrogen ions and electrons forming water.

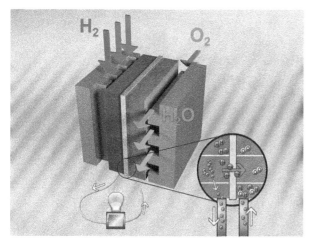

Figure 3.7.23: *Fuel cells provide a clean and efficient power source*

The ion-electron equations for the oxidation and reduction half-reactions in the cell are:

$$H_2(g) \rightarrow 2H^+(aq) + 2e^- \quad \textit{oxidation}$$

$$O_2(g) + 4H^+(aq) + 4e^- \rightarrow 2H_2O(\ell) \quad \textit{reduction}$$

The overall equation for the redox reaction is:

$$2H_2(g) + O_2(g) \rightarrow 2H_2O(\ell) \quad \textit{redox}$$

Using fuel cells will help reduce carbon dioxide emissions even when the hydrogen used as fuel is made from fossil fuels. This is because using fuel cells is much more efficient than burning fuels.

Figure 3.7.24: *Fuel cells are increasingly being used in place of internal combustion engines for transport and in industry for machines such as forklifts*

◄Ε SPOTLIGHT ON THE ENVIRONMENT

Until 2007 only about 2% of batteries were recycled. The remaining 98% went into landfill.

Many batteries are made using heavy metals such as nickel, cadmium, mercury and lead. Heavy metals are toxic and disposal of batteries into landfill has the potential to cause environmental problems. When the outer casings of the batteries corrode, the materials inside can dissolve and leach out from the landfill site into the surrounding environment – including rivers and streams – causing soil and water pollution. Cadmium from discarded batteries has been detected in the world's oceans. If batteries are incinerated with household waste the heavy metals will get into the atmosphere, increasing air pollution.

Figure 3.7.25: *Batteries contain heavy metals, making them hazardous waste*

A European Union directive came into force in the UK in 2010. This stated that by 2012, 25% of batteries should be recycled and that by 2016, 45% must be collected and recycled. Retail stores have been required by EU law to set up collection points where used batteries can be recycled.

Recycling doesn't only help deal with potential pollution problems, many of the materials in the batteries can be recovered and re-used, preserving precious resources.

Learning checklist

In this chapter you have learned:

- In metals the outer electrons are delocalised.
- Metallic bonding involves the attraction of metal nuclei for delocalised outer electrons.
- The properties of metals, including electrical conductivity, are explained by metallic bonding.
- Metals can be arranged in order of reactivity by comparing the rates at which they react with oxygen, water and acids.
- Some insoluble salts can be produced when metals react with acids.
- The different methods of obtaining metals from their ores include heating the ore, smelting the ore and electrical methods.
- Extraction of the metal from its ore is called reduction.
- Reduction of ores involves metal ions gaining electrons.
- Oxidation is the opposite of reduction and involves the loss of electrons by an atom or ion.
- Reduction and oxidation always take place together. The combined reaction is called a redox reaction.
- Ion-electron equations can be written to describe the processes of reduction and oxidation.
- Redox equations can be formed by combining the ion-electron equations for reduction and oxidation.
- The reactions of metals with water, oxygen and acids are examples of redox reactions and can be described using redox equations.
- Electrically conducting solutions containing ions are known as electrolytes.
- A simple electrochemical cell can be made by connecting metals in an electrolyte.
- Electrochemical cells can be made by connecting metals dipped in solutions of their own ions, connected by an ionic (salt) bridge.
- Different pairs of metals produce different voltages, which can be used to arrange the metals into an electrochemical series.

- The further apart the metals are in the electrochemical series, the greater the voltage produced in an electrochemical cell.

- In an electrochemical cell, the electrons flow from the higher metal to the lower, through the connecting wire.

- Electricity can be produced in a cell where the electrodes are not metal.

- Graphite (carbon) electrodes are used in chemical cells because they conduct electricity and do not react with the electrolytes.

- The reactions in electrochemical cells are examples of redox reactions.

- The reactions at the electrodes in electrochemical cells can be described using ion-electron equations.

- Ion-electron equations for the reactions at the electrodes can be combined to give a redox equation for the cell reaction.

8 Plastics

SPOTLIGHT ON SPORT: LONDON 2012 – THE 'PLASTIC' OLYMPICS

With its pink fringe and electric blue surface, the London Olympics hockey pitch certainly caught the attention of the world – it didn't take long for the players to nickname it the 'Smurf Turf'. Not only did it look striking, it performed to the high standards required – it had to have a smooth surface but also be tough enough to take the wear from players' boot studs. The artificial grass pitch was made from a type of poly(ethene) specially developed for use at the Olympics. It wasn't just the surface itself that was plastic; the base for the pitch was made from synthetic rubber and the pitch was glued to the base using a polyurethane adhesive.

Figure 3.8.1: *The 'Smurf Turf' – the London 2012 Olympic hockey pitch*

The ability of the plastic to absorb and retain colour was important to those watching in the stadium and on TV because the yellow ball that was used showed up better on the blue pitch than it did on the traditional green. At the end of the games the pitch was rolled up and moved to another part of the Olympic park for public use.

The Olympic shooting venue comprised of three white polyvinylchloride (PVC) tents which allowed natural light through. They were temporary buildings so that they could be reused – they were dismantled immediately after the Olympics to be used in Glasgow for the 2014 Commonwealth Games. All of the temporary structures were dismantled after the Olympics and either recycled or reused. Some PVC panels used in the structure of the main stadium were kept and used in football stadiums built for the 2014 FIFA World Cup in Brazil.

Figure 3.8.2: *One of the white PVC tents at the London 2012 Olympic shooting venue*

The Aquatics centre used over 8000 square metres of PVC for its external wrap. In the Velodrome PVC was used on its high performance surface. The main Olympic stadium itself made extensive use of plastics on the athletics track and in the cladding which was wrapped around the stadium. There were 336 giant individual panels made from polyester and polythene.

The manufacturer of the US team's uniforms for the games claims that an average of 82% recycled polyester fabric and up to 13 recycled plastic bottles were used per uniform. The company also made football boots worn by many Olympians – 70% of the boot was made from recycled plastic including polyester from water bottles and other packaging.

Figure 3.8.3: *The outfits worn by most athletes at the London 2012 Olympics had a high recycled plastic content*

The whitewater kayaking course was constructed using plastic blocks instead of concrete. The blocks were very light so the course could easily be reconfigured to alter the flow of water. This also meant that after the Olympics the blocks could be moved around to make the course more suitable for recreational use.

It was not just the stadiums and pitches which featured plastics; much of the equipment used by the competitors was made of plastic. The kayaks used in the whitewater kayaking and the helmets worn were both made from lightweight but tough plastics. The frames of the bicycles used in the Velodrome and in the road race are no longer made from aluminium and steel. Almost all bike frames are now made from carbon fibre-reinforced plastic (CFRP), which combines extremely light weight with rigidity. Professional-level racing bike frames can weigh as little as 700 g – less than a bag of sugar.

Spectators at the games played their part in the development of plastics. The plastic bottles containing the beer sold at the venues were specially developed for use at the games. Not only were they made from the recyclable polyester PET, they were made with new technology that greatly reduced the rate that the gas in the drink escaped through the plastic.

Figure 3.8.4: *A carbon fibre-reinforced plastic (CFRP) bike frame used at the London 2012 Olympics weighed less than a bag of sugar*

Activity 3.8.1

Make up a table showing how different plastics were used at the London 2012 Olympics.

Addition polymerisation

The poly(ethene) used to make the playing surface of the London 2012 Olympic hockey pitch is an example of an addition polymer. Addition polymers are formed when lots of small unsaturated molecules (monomers) join together – in the case of poly(ethene) the monomer is ethene. The characteristic which makes hydrocarbon molecules like ethene suitable to undergo an addition reaction is the carbon to carbon double bond (C=C). Under the right conditions the double bond breaks, resulting in electrons being made available to bond with neighbouring ethene molecules. Although the diagram only shows three molecules joining, in practice thousands of molecules join together.

Hint

Plastics are examples of polymers.

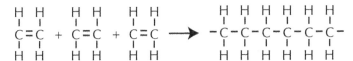

Figure 3.8.5: *Ethene mwolecules add together when the double bonds break*

Chemists can vary the chain length and the number and size of the branches in a polymer in order to change its properties. We are all familiar with 'polythene' bags used for carrying shopping and general packaging – polythene is the common name for poly(ethene). This type of polythene is known as low-density poly(ethene) (LDPE) – it is soft and very pliable, which makes it useful for packaging and electrical insulation. LDPE molecules contain about 50 000 carbon atoms which have a lot of branching, which results in LDPE being soft. Another form of poly(ethene) is high-density poly(ethene) (HDPE). HDPE molecules contain more than 50 000 carbon atoms with little branching of the main chains, resulting in more rigid materials with a higher range of melting points.

Activity 3.8.2: Paired activity

Work with a partner to do this activity.

1. Collect a box of molecular models and each make an ethene molecule.
2. Break apart one of the bonds which make up the C=C double bond.
3. Join the two molecules together – note the double bond is no longer there.
4. Make more monomers and add them to your structure – if others in the class are doing the same exercise join your structures together.
5. You will end up with part of the poly(ethene) structure – note that both ends of the structure are open and could react further.

Figure 3.8.6: *The bottle on the left is made from HDPE and the bottle on the right is made from LDPE*

The surface of the hockey pitch used in the London 2012 Olympics was made from yet another form of poly(ethene) called linear low-density poly(ethene) (LLDPE). Using octene as a monomer the poly(ethene) produced was found to be up to 30% more durable than LDPE but still maintained the softness required for the surface. The artificial turf used for the hockey pitch was tested by sliding two studded rollers over a sample piece of turf over 20 000 times and still no significant wear was observed.

The structure of monomers and polymers

Poly(propene) is another widely used addition polymer. It is very tough and flexible and is used as an engineering plastic. Since the 1990s a number of countries have been using a form of poly(propene) for making banknotes. It is a very durable polymer so the banknotes have a longer lifetime compared to paper. It is also possible to incorporate many security features on the notes to make counterfeiting more difficult, which can't be done with paper notes.

The propene molecules add together in a similar way to the ethene molecules when poly(ethene) is produced:

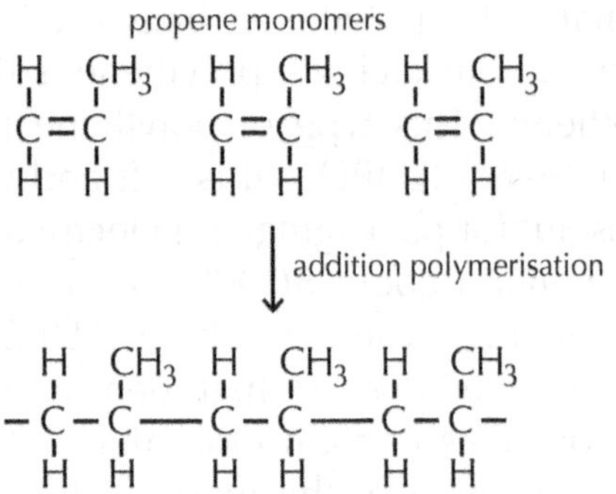

propene monomers

$$\begin{array}{ccc} H \quad CH_3 & H \quad CH_3 & H \quad CH_3 \\ C=C & C=C & C=C \\ H \quad H & H \quad H & H \quad H \end{array}$$

↓ addition polymerisation

$$\begin{array}{ccc} H \quad CH_3 & H \quad CH_3 & H \quad CH_3 \\ -C-C & C-C & C-C- \\ H \quad H & H \quad H & H \quad H \end{array}$$

poly(propene)

Figure 3.8.7: *Propene molecules adding to make poly(propene)*

Note that the structure of the monomer is drawn in an ⊢⊣ shape so it is easy to see how the monomers join – two ways of drawing a propene molecule are illustrated here, clearly showing the ⊢⊣ shape of the preferred structure.

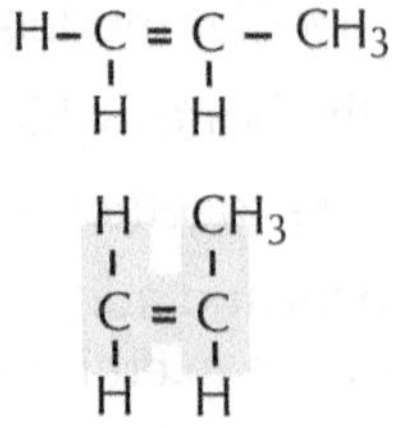

$$H-C=C-CH_3$$
$$\begin{array}{cc} H & H \end{array}$$

$$\begin{array}{cc} H & CH_3 \\ C=C \\ H & H \end{array}$$

The repeating unit in an addition polymer can be identified from the structure of the polymer. The structure shown is part of the perspex polymer:

$$\cdots \begin{matrix} H & CH_3 \\ | & | \\ -C-C- \\ | & | \\ H & COOCH_3 \end{matrix} \begin{matrix} H & CH_3 \\ | & | \\ C-C \\ | & | \\ H & COOCH_3 \end{matrix} \begin{matrix} H & CH_3 \\ | & | \\ C-C- \\ | & | \\ H & COOCH_3 \end{matrix} \cdots$$

The broken lines highlight the unit which repeats in the structure:

$$\begin{matrix} H & CH_3 \\ | & | \\ -C-C- \\ | & | \\ H & COOCH_3 \end{matrix}$$

Notice that the end bonds in the repeating unit are left open. The structure of the monomer can be worked out from the repeating unit:

$$\begin{matrix} H & CH_3 \\ | & | \\ C=C \\ | & | \\ H & COOCH_3 \end{matrix}$$

The monomer must always have a C=C double bond.

Naming polymers

Addition polymers are named from the monomer used. For example, the monomer chlorethene is used to make poly(ethene). Table 3.8.1 shows the names and uses of some common addition polymers.

Table 3.8.1: *Names and uses of some common addition polymers*

Alkene monomer	Structural formula	Polymer	Structure of polymer (repeating unit)	Uses
ethene	$\begin{matrix} H & & H \\ & C=C & \\ H & & H \end{matrix}$	poly(ethene)	$\left[\begin{matrix} H & H \\ | & | \\ C-C \\ | & | \\ H & H \end{matrix}\right]$	bags, insulation for wires, squeezy bottles
propene	$\begin{matrix} H & & CH_3 \\ & C=C & \\ H & & H \end{matrix}$	poly(propene)	$\left[\begin{matrix} H & CH_3 \\ | & | \\ C-C \\ | & | \\ H & H \end{matrix}\right]$	bottles, plastic plates, clothing, carpets, crates, ropes and twine
phenylethene	$\begin{matrix} H & & C_6H_5 \\ & C=C & \\ H & & H \end{matrix}$	poly(phenylethene) (polystyrene)	$\left[\begin{matrix} H & C_6H_5 \\ | & | \\ C-C \\ | & | \\ H & H \end{matrix}\right]$	insulation, packaging, food containers, model kits, flowerpots, housewares
tetrafluoroethene	$\begin{matrix} F & & F \\ & C=C & \\ F & & F \end{matrix}$	poly(tetrafluoroethene) (Teflon®, PTFE)	$\left[\begin{matrix} F & F \\ | & | \\ C-C \\ | & | \\ F & F \end{matrix}\right]$	non-stick pans, lubricant-free bearings

Activity 3.8.3

The structure for part of the poly(chloroethene) polymer chain is:

$$-\overset{\displaystyle H}{\underset{\displaystyle H}{C}}-\overset{\displaystyle Cl}{\underset{\displaystyle H}{C}}-\overset{\displaystyle H}{\underset{\displaystyle H}{C}}-\overset{\displaystyle Cl}{\underset{\displaystyle H}{C}}-\overset{\displaystyle H}{\underset{\displaystyle H}{C}}-\overset{\displaystyle Cl}{\underset{\displaystyle H}{C}}-$$

(a) Draw the structure of its repeating unit.
(b) Draw the structure of the monomer unit.
(c) Name the monomer unit.
(d) Find out the common name for poly(chloroethene).

SPOTLIGHT ON INDUSTRY

The uses of HDPE relate to its different properties. It is thermosoftening, chemically very unreactive and retains its shape well at higher temperatures than LDPE. It can be blow moulded to make bottles and other containers for a range of products from shampoo to household cleaners and bleach. During blow moulding the HDPE is softened by heating it, compressed gas is blown in and the plastic forms the shape of the mould. Hospital appliances such as bed pans are made from HDPE because its higher melting point means they can be sterilised at higher temperatures without losing their shape.

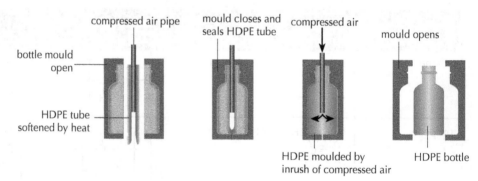

Figure 3.8.8: *Blow moulding a HDPE bottle*

Almost half of the HDPE produced is used in blow moulding. Injection moulding accounts for the rest of the HDPE produced. Again the plastic is softened and injected into a mould of the required shape and allowed to harden. Food storage containers, buckets and crates are made this way.

STEP BACK IN TIME: NYLON

Nylon was first made in a laboratory of the Du Pont company in the USA in the 1930s, by a group of chemists led by Wallace Carothers who was the leading expert in plastic chemistry at the time. It was quickly realised that nylon had useful properties: it was exceptionally strong, easily made into a fibre and could be made to look very shiny. The first commercial use of nylon was as the bristles in a toothbrush. Up until then the bristles had been made from animal bristles. It was soon realised that there was great potential to use nylon to make women's stockings. At that time stockings were made from silk, which was expensive. Nylon was cheap and the sheen produced when it was made into a fibre made it a good substitute for silk. Nylon stockings became widely available in 1940. However, by that time World War II had started and nylon was needed to make ropes and, in particular, parachutes, which up until then had been made from silk. Nylon stockings were rationed until the end of the war and it was not until the 1950s that they became widely available in Western Europe. Today, nylon goods account for 95% of the women's hosiery market.

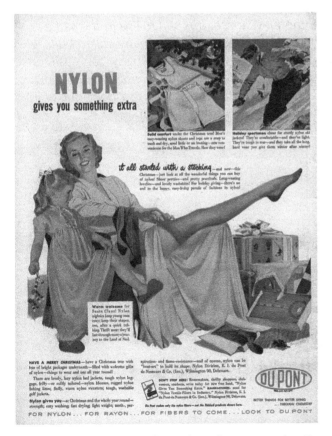

Figure 3.8.9: *1940s advert for stockings known as 'nylons'!*

Figure 3.8.10: *Tapping latex from a rubber tree, Thailand*

SPOTLIGHT ON NATURAL POLYMERS

Natural rubber is a mixture of polyisoprene and small amounts of other chemicals including water. Rubber is often referred to as an elastomer (elastic polymer). Rubber is derived from latex, a milky substance produced by some plants. The main source of latex is the rubber tree, which is mostly cultivated in parts of Asia. The trees are 'tapped' by making an incision into the bark of the tree and the sticky latex sap collected and refined into a usable rubber. It is normally very stretchy and flexible and extremely waterproof. The ability to stretch is due to the shape of the polymer chains. In the relaxed form the chains are tangled but when stretched they align with each other and then relax when the pressure is released. Most of the natural rubber produced is used in making pneumatic tyres for road vehicles.

An example of an item made from 100% natural rubber is the baby teething toy called 'Sophie la girafe'. Sophie was first made in France in 1961 from natural rubber derived from the sap of the Hevea tree. It is still made in the same way using 100% natural rubber by a company called Vulli. The toy is shaped in such a way that the baby can easily hold and bite on it and it is soft enough not to hurt the baby's mouth. The toy can be stretched and bent out of shape but quickly returns to its original shape.

Figure 3.8.11: *Sophie la girafe$^{(TM)}$, Vulli SAS*

Natural polymers are biodegradable – they can be broken down by microorganisms in the soil and air.

Bioplastics are synthetic and are a form of plastic derived from renewable biomass sources, such as vegetable fats and oils, corn starch and pea starch. They were the basis for the production of the earliest plastics and are making a comeback. The monomers used to make common plastics like poly(ethene) are mainly derived from oil fractions, which are becoming scarce – oil is a finite resource. There are a variety of bioplastics being made – some common uses are as packaging materials and insulation.

Figure 3.8.12: *Packaging made from corn starch is similar to expanded polysryrene but is biodegradable*

Starch-based plastics are the most widely used bioplastic. Like starch itself they can absorb water and are used for making drug capsules. Industrially, starch-based bioplastics are often blended with biodegradable polyesters which make them compostable. Cellulose-based bioplastics such as cellulose acetate are used to make transparent food packaging. Many bioplastics do not degrade, a property which is useful but environmentally a problem when it comes to disposing of the plastic.

Poly(lactic acid), known as PLA, is a bioplastic which is degradable. It is made from corn or sugar cane. Bacteria ferment the sugar to produce lactic acid which is then polymerised. PLA has been used to make disposable cutlery, waste sacks and sutures used for internal stitches in the body. You may have drunk tea made from tea enclosed in a PLA tea bag. PLA is both biodegradable and photodegradable – it absorbs UV radiation, which causes the polymer to break down.

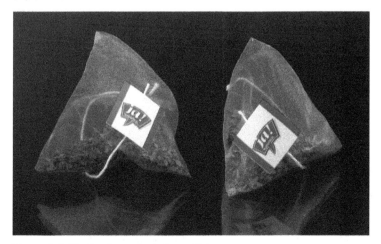

Figure 3.8.13: *Tea bags made from PLA*

Polyhydroxybutyrate (PHB) is a polyester that has been 'grown' using bacteria to ferment glucose – it is better known by its trade name Biopol. The plastic produced solidifies into a polymer similar to poly(ethene), but is fully biodegradable. PHB is produced naturally in the stomach, so is not rejected by the body, which opens up the possibility of it being used as a coating for slow-release drugs. A major problem with its use is that its production costs are very high.

Synthetic plastics can be made biodegradable in a number of ways. Natural polymers such as starch, cellulose and protein can be impregnated into plastic items like polythene bags. If the bag is dumped in a landfill site, microorganisms in the soil quickly break down the natural material in the polythene and it disintegrates.

Figure 3.8.14: *The bioplastic cutlery, coffee cup and lid are made from PLA and are suitable for commercial composting together with food waste*

◄◄ SPOTLIGHT ON NOVEL PLASTICS

Figure 3.8.15: *Kevlar is used in bullet-resistant vests (above) because it is tough and lightweight*

Some plastics have been discovered by accident and others have been known for some time before they were put to use. What is certain is that chemists are continually looking to improve the properties of plastics in order that they can be used more and more in everyday life. Plastics with unusual properties are often called novel plastics.

Kevlar is a plastic that had been discovered years before a suitable solvent was found to dissolve it and make it usable. Kevlar polymer chains can be aligned to make very strong fibres which can be twisted to make cables that have the same strength as steel cables but are five times lighter. It can be used in vehicle tyres instead of steel to improve the strength of the rubber and make the tyre lighter. Kevlar is well known as the protective padding in bullet-resistant vests. It is also used in some aircraft wings where strength combined with very little weight is essential.

Polyvinyl alcohol (PVA) is a soluble plastic which has a huge number of uses. Solubility is not a property usually associated with plastics but PVA combines solubility with strength, which makes it useful for making laundry bags for use in hospitals, where infection may be a risk – the bag dissolves as the linen is washed and rinsed. Some detergents for use in dishwashers and washing machines are wrapped in PVA which quickly dissolves, releasing the detergent.

PVA itself is used to make a polymer fibre known commercially as Vinalon. It is the most widely used fibre in the Democratic People's Republic of Korea (North Korea) despite it having inferior properties to other synthetic fibres. This is because the country prefers to be self-sufficient and the manufacture of Vinalon doesn't require oil as a source of raw material so the country doesn't have to import it from other countries.

Sodium polyacrylate is a polymer which absorbs water. Generally plastics repel water, which is one of their most useful properties. Chemists however saw the potential for using sodium polyacrylate gel beads (hydrogels) in disposable nappies. Hydrogels can absorb 200–300 times their own weight of water. One problem using them in nappies is that the hydrogel absorbs the water too quickly, which causes clogging, which in turn cuts down the efficiency of the gel. The problem was solved by adding natural materials, such as wood pulp, which quickly absorb the water then slowly disperse it through the sodium polyacrylate. Traditional plastics are also used in disposable nappies to stop liquid leaking out. Chemicals which change colour when they are wet are sometimes added so that it can easily be seen if the nappy is leaking.

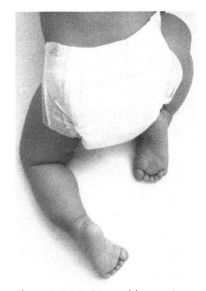

Figure 3.8.16: Disposable nappies contain hydrogel which absorbs water

A group of scientists spent 17 years developing a plastic marketed as RhinoPlex which is used to instantly repair tyre punctures. The plastic is sprayed onto the inside of the tyre and if it is punctured the plastic immediately fills the hole and so stops the air from leaking out. It is also claimed that using the plastic helps stabilise the vehicle, thus saving energy, cuts road noise and extends the tyre's life. One of the downsides of this plastic is that its use increases the weight of the tyre, which in turn uses up more fuel.

'Smart' plastics which react to their environment are being developed. Chemists in the USA have developed a self-healing plastic which mimics the skin's ability to heal scratches and cuts. It could potentially be used as a self-repairing

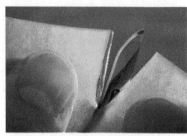

Figure 3.8.17: *A self-healing plastic being cut*

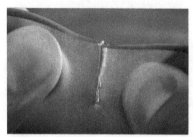

Figure 3.8.18: *The self-healing plastic repairs itself when held together*

surface for the likes of mobile phones and laptops. The plastic changes colour, turning red to warn that there is damage to the surface, and will then heal itself when exposed to light. These 'smart' plastics have great potential for use in areas where safety is a major issue such as in aircraft wings and bridges. If a crack developed it would turn red so it could be easily spotted. Engineers could then investigate and decide whether to repair or replace the damaged area.

The incorporation of colour-changing plastics into food packaging materials could be used to let consumers know the conditions inside a food package. Plastics which are extremely sensitive to stress have been developed. The transparent plastic is stretched across the food and if the food starts to go off gases are given off which cause bloating of the plastic film. This causes the molecules in the plastic film to stretch out and they interact with light in such a way as to cause colour to appear in the plastic. The appearance of colour would let the customer know that there could be a problem with the food. This colour-changing technology could replace coloured dyes as indicators of food quality.

Antioxidants are important chemicals which slow down the reaction of oxygen in the air with food. Chemists at a US Army research facility have been developing a process to improve natural antioxidants like the ones found in olives. They have been able to improve the power of the antioxidant by polymerising it to give a longer molecule. The more powerful the antioxidant, the longer the shelf-life of the food.

Plastics usually conduct electricity so poorly that they are used to insulate electric cables. Australian researchers have shown that by placing a thin film of metal onto a plastic sheet and mixing the metal into the plastic using an ion beam they can make cheap, strong, flexible and conductive plastic films. It is predicted that this discovery has the ability to open up new avenues for making plastic electronics. In theory plastics could be made that conduct no electricity at all or as well as metals do – and everything in between. Conductive plastics could make flexible touch screens and e-paper a realistic possibility in the near future.

Learning checklist

In this chapter you have learned:

- Plastics are examples of polymers.

- Polymers are long chained molecules formed by joining a large number of small molecules called monomers.

- Addition polymers are formed from unsaturated monomers.

- The name of the addition polymer comes from the name of the monomer.

- The repeating unit is the shortest section of a polymer chain.

- The structure of an addition polymer can be drawn from the structure of its monomer or the repeating unit.

- The structure of the repeating unit and the monomer can be drawn from the structure of an addition polymer.

- An addition polymer can be recognised from its carbon backbone.

9 Fertilisers

You should already know

- Plants help to sustain life by providing food and oxygen.
- Chemists have an important role in ensuring there is sufficient food production.
- Plants need nutrients for healthy growth.
- Nutrients can be provided by fertilisers.
- Natural fertilisers include compost and manure.
- Overuse of synthetic fertilisers can cause environmental problems.

Learning intentions

In this chapter you will learn about:

- Commercial production of fertilisers.
- The Haber process – the industrial manufacture of ammonia.
- The Ostwald process – the industrial manufacture of nitric acid.

Commercial production of fertilisers

In August 2012 the United Nations estimated that the world population will rise from 7 billion to 10·1 billion by the end of the century. One of the main consequences of this is the challenge of providing the provision of food for the world.

Table 3.9.1: *World population estimates according to the United States Census Bureau (USCB)*

Population (in billions)	1	2	3	4	5	6	7	8	9
Year	1804	1927	1960	1974	1987	1999	2012	*2027*	(x)
Years elapsed	—	123	33	14	13	12	13	*15*	(y)

Activity 3.9.1

(a) Use the figures in the world population estimates table to draw a line graph of population against year. Remember to label your axes clearly.

(b) From your graph, predict in which year the world population will reach 9 billion (x) and then work out years elapsed (y).

(c) Use the information in the 'years elapsed' part of the table to make a general statement about the rate of population growth between 1804 and (x).

All of our food relies on plant growth – either for eating the plants directly or eating animals which have eaten plants. Food crops for animals and humans rely on nutrients to supply the essential elements for plant growth. These essential elements are nitrogen, phosphorus and potassium (N, P and K). Substances that are added to the soil to restore the nutrients taken out by plants are called **fertilisers**.

📖 **Word bank**

• **Fertilisers**

Substances that are added to the soil to restore the nutrients taken out by plants.

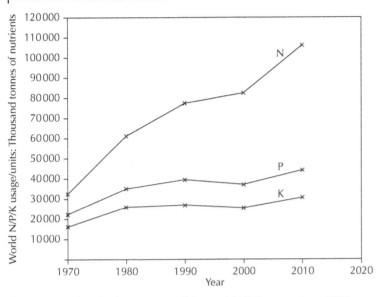

Figure 3.9.1: *Graph showing growth in world NPK usage since 1970*

Source: International Fertilizer Industry Association (IFA)

GO! Activity 3.9.2

Use the information from the graph showing the growth in world NPK usage to answer the questions.

(a) Why do you think the amount of NPK being used is generally increasing every year?

(b) Estimate how many millions of tonnes of N, P and K were used worldwide in 2015.

(c) Predict what the usage trend will be through the rest of the century and give a reason for your prediction.

Feeding the world is not a new problem. At the beginning of the last century there was worldwide concern that natural nitrate compounds, a source of nitrogen essential for plant growth, would run out. Most of the world's natural nitrates came from South America. The vast cliffs of bird droppings, known as 'guano', found in Peru, had run out and although there were still huge deposits of nitrates in Chile they had to be transported across the world, which made them very expensive. The need for fertilisers was not the only reason nitrates were so sought after. Nitrates were needed to make explosives for use in mining and for making armaments – this was the beginning of the twentieth century and by 1914 World War I had begun.

The air is made up of nearly 80% nitrogen and at first sight it may seem an obvious source of nitrogen for plants. Unfortunately only plants with bacteria containing nodules on their roots can take the nitrogen from the air and convert it to soluble nitrates which can be absorbed by the plant. Nitrogen is a very unreactive element and although chemists had managed to make compounds of nitrogen they had not managed to find cost-effective routes to produce them on an industrial scale. In 1909 a German chemist called Fritz Haber managed to make ammonia (NH_3) from nitrogen and hydrogen and this was the first step towards an economical route to synthetic fertilisers.

Making ammonia in the lab

Ammonia can be made in the laboratory by heating an ammonium salt, such as ammonium chloride, with a base like solid soda lime or sodium hydroxide solution:

$$NH_4Cl(s) + NaOH(aq) \rightarrow NaCl(aq) + H_2O(l) + NH3(g)$$

Notice in Figure 3.9.2 that the moist pH paper turns blue, showing that ammonia is alkaline.

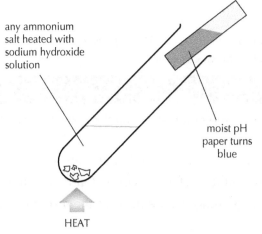

Figure 3.9.2: *Heating an ammonium salt and sodium hydroxide solution to produce ammonia*

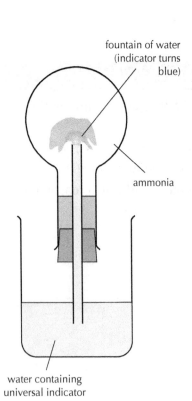

Figure 3.9.3: *The fountain experiment*

Ammonia is a colourless gas which is extremely soluble in water. Your teacher may demonstrate this by doing the fountain experiment (figure 3.9.3). A dry flask filled with ammonia gas is inverted in a beaker of water. The ammonia is so soluble that it quickly starts to dissolve and the water from the beaker is drawn into the flask like a fountain. The universal indicator turns blue showing that an alkaline solution has been made.

Ammonia has a very strong smell which can affect your breathing – it is the smell often associated with wet nappies.

Ammonia and ammonia solution reacts with acids to form soluble salts which can be used as fertilisers. Ammonium nitrate, one of the most widely used nitrogen fertilisers, is produced when ammonia reacts with nitric acid.

$$\text{ammonia + nitric acid} \rightarrow \text{ammonium nitrate}$$

$$NH_3(g) + HNO_3(aq) \rightarrow NH_4NO_3(aq)$$

The Haber process – the industrial manufacture of ammonia

The Haber process is the basis for making ammonia industrially, from nitrogen and hydrogen.

The balanced equation for the reaction is:

$$N_2(g) + 3H_2(g) \rightleftharpoons 2NH_3(g)$$

The $\rightleftharpoons$ symbol indicates that the reaction is reversible. This means that as the ammonia is produced, some of it breaks down and reforms the reactant.

The reaction conditions need to be controlled to get as much ammonia as economically as possible.

Most manufacturers use the following conditions:

- An average temperature of 450°C. This is a compromise temperature at which the rate of reaction is fast enough to give a reasonable amount (yield) of ammonia in a short time.

- A pressure of around 200 atmospheres. Higher pressures are very costly in terms of running the plant and also more expensive to construct, as thicker walled pipes are needed.

- A finely divided iron catalyst to speed up the reaction. Using finely divided iron means the catalyst has a bigger reacting surface.

- The ammonia is usually liquefied as it forms and drained off.

These conditions provide an average ammonia yield of around 25% in a single run. The unreacted nitrogen and hydrogen are recycled, which pushes up the yield to 98%. Recycling makes the process more cost effective because valuable resources are not wasted and avoids the release of hydrogen into the atmosphere. Chemists have been developing new catalysts that operate at lower temperatures and pressures in an attempt to lower the energy costs of making the ammonia, which in turn saves valuable finite fuels.

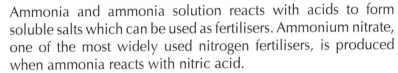

Hint

Don't get confused between ammonia and ammonium.

Ammonia is the gas made in the Haber process and has the formula NH_3.

Ammonium is the ion with the formula NH_4^+. The ammonium ion is always part of a compound, e.g. ammonium nitrate.

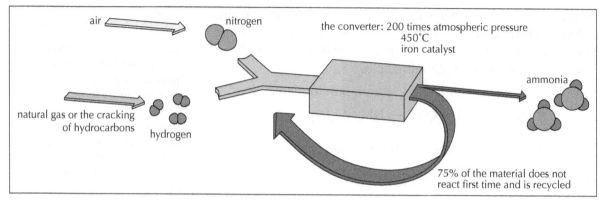

Figure 3.9.4: *Flow diagram summarising the Haber process*

🚲 STEP BACK IN TIME: THE WORK OF FRITZ HABER

The balanced equation for the reaction of nitrogen with hydrogen to produce ammonia is:

$$N_2(g) \ + \ 3H_2(g) \ \rightleftharpoons \ 2NH_3(g)$$

The $\rightleftharpoons$ symbol indicates that the reaction is reversible, which means that as the nitrogen and hydrogen react to form ammonia some of the ammonia breaks down and reforms the reactants, i.e. the reaction goes in a forward direction and a backward direction. Haber's research focused on forcing the reaction to go forward so that more ammonia formed than broke down.

It was well known that changing the temperature affects the rate of reactions but Haber found that too high a temperature caused the ammonia to break down. However, too low a temperature slowed the rate at which the ammonia formed.

Haber also knew that because the reactants and products were gases changing the pressure would have an effect on the amount of ammonia produced. The equipment he used could only raise the pressure to 200 atmospheres but that was high enough to increase the amount of ammonia obtained.

Haber carried out more than 6000 experiments to try and find a catalyst that would help produce the ammonia faster. He eventually found that iron gave the best results.

In 1913 a chemical engineer called Carl Bosch scaled up Haber's laboratory arrangement for making ammonia and made the first industrial ammonia plant. Haber and Bosch were awarded a Nobel Prize in 1918 for their work in the production of ammonia.

Figure 3.9.5: *Fritz Haber (left) with Albert Einstein. Haber won the 1918 Nobel Prize in Chemistry*

◄€ SPOTLIGHT ON PRODUCTION AND THE USES OF AMMONIA

Countries which make ammonia usually have a ready supply of natural gas. The gas, which is mainly methane, undergoes a process called steam reforming to produce hydrogen. The nitrogen required is mainly obtained by low-temperature distillation of the air. The natural gas and the air are the feedstocks for the process.

Most of the ammonia produced in the world is used to make fertilisers. Other uses of ammonia are:

- *in the manufacture of nitric acid (HNO_3)*
- *in the manufacture of nylon and other polyamides*
- *as a refrigerant in household, commercial and industrial refrigeration systems*
- *in the manufacture of dyes and explosives*
- *as an active ingredient in some cleaning solutions.*

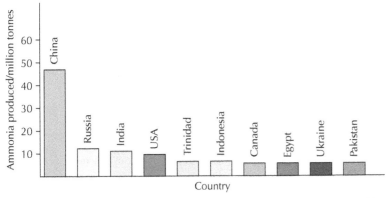

Figure 3.9.6: *Graph showing the top ten ammonia producers in the world*
Source: International Fertilizer Industry Association (IFA)

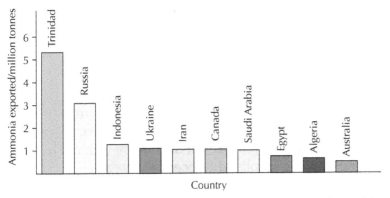

Figure 3.9.7: *Graph showing the top ten ammonia exporters in the world*
Source: International Fertilizer Industry Association (IFA)

GO! Activity 3.9.3

Use the information in Figures 3.9.6 and 3.9.7 to help you.

1. (a) Where does China rank in terms of world ammonia production?

   (b) From the graph of world exporters what can you say about China in terms of how much ammonia it exports?

   (c) Suggest why there is this difference.

2. (a) Where does Trinidad rank in terms of ammonia exporters?

   (b) Which natural resource is Trinidad likely to have lots of in order to make ammonia?

   (c) Look at a map of the world, find Trinidad and suggest which major country is likely to import their ammonia from Trinidad. Give a reason for your answer.

3. This graph shows the effect of pressure and temperature on the conversion of nitrogen and hydrogen to ammonia in the Haber process.

   (a) At what temperature is the conversion rate the highest?

   (b) What happens to the conversion rate when the pressure is increased?

   (c) Use the graph to find the conversion rate when the temperature is 400°C, with a pressure of 500 atmospheres.

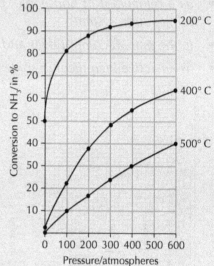

The Ostwald process – the manufacture of nitric acid

Some of the ammonia produced by the Haber process is used to make nitric acid (HNO_3). It might seem more obvious to try to combine nitrogen and oxygen from the air to form nitrogen dioxide which then dissolves in water to give nitric acid. Although it seems straightforward, the problem of nitrogen's lack of reactivity means that huge amounts of electricity are needed to supply the energy required to make the nitrogen react. Nitrogen and oxygen do react in the air but only during lightning storms when the lightning supplies the necessary energy.

Wilhelm Ostwald is credited with inventing the industrial method of converting ammonia into nitric acid, although the chemistry of the process had been known for some time. Ostwald had experimented for many years with catalysts and it was his knowledge in this area of chemistry that proved to be key. He found that if a mixture of ammonia and air was passed over a heated platinum catalyst, nitrogen dioxide was formed which dissolved in water to form nitric acid. The method he patented in 1902 is basically the same as is used today. The development of the Haber process for the production of ammonia allowed the large scale production of nitric acid to begin.

The following flow diagram gives a simplified outline of the Ostwald process.

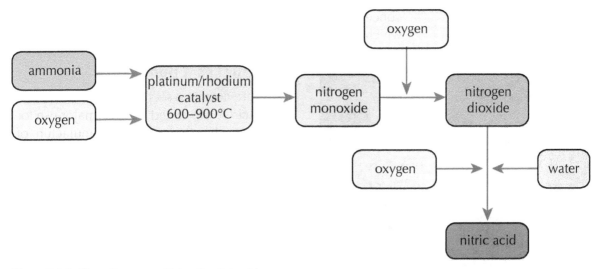

Figure 3.9.8: *Flow diagram outlining the Ostwald process*

The first step is the catalytic oxidation of ammonia. Ammonia and air are passed through several layers of platinum/rhodium gauze which can be several metres in diameter. This provides a large surface area for the reaction to take place on. The catalyst is heated to between 600–900°C. The reaction is exothermic. Once the catalyst is heated it will remain at a high temperature so the external source of heat can be switched off, which saves money. Although increasing the pressure improves the yield, it is not a lot of difference compared to atmospheric pressure so the pressure is only raised slightly. Not having to use high pressure is another cost saving.

ammonia + oxygen → nitrogen monoxide + water

$$4NH_3(g) + 5O_2(g) \rightarrow 4NO(g) + 6H_2O(g)$$

In the second step the nitrogen monoxide reacts with more oxygen forming nitrogen dioxide, a distinctive brown gas:

nitrogen monoxide + oxygen → nitrogen dioxide

$$2NO(g) + O_2(g) \rightarrow 2NO_2(g)$$

In the final step nitrogen dioxide is reacted with water and more oxygen to form nitric acid:

nitrogen dioxide + oxygen + water → nitric acid

$$4NO_2(g) + O_2(g) + 2H_2O(\ell) \rightarrow 4HNO_3(aq)$$

In industry the nitrogen dioxide and oxygen are passed up an absorption tower which is packed with glass beads over which water flows. The gas mixture and water meet at the surface of the beads and react to form nitric acid. The acid collected at the bottom of the tower is 68% nitric acid and 32% water. The acid is concentrated by separating the water during a distillation process which involves concentrated sulphuric acid which absorbs water.

Figure 3.9.9: *Two bags of NPK fertiliser each showing different proportions of N, P and K*

What makes a good fertiliser?

In the previous two sections on ammonia and nitric acid it was stated that the major use for both these chemicals is in the manufacture of fertiliser. A good fertiliser must contain at least one of the essential elements (N, P and K) and be soluble in water so that it can be absorbed through the roots of the plant. Table 3.9.2 lists compounds which are suitable for use as fertilisers. The table includes the percentage by mass of each of the essential elements. Notice that no one compound contains all of the essential elements. Most plants require at least some of each essential element so fertilisers are usually mixtures containing varying amounts of the compounds containing the essential elements. They are often referred to as NPK fertilisers.

Calculating the percentage mass composition of nitrogen in fertilisers

Table 3.9.2 shows the percentage composition by mass of the essential elements N, P and K found in common fertilisers. This can easily be calculated from the relative atomic masses (RAM) of the elements in the compound as shown for ammonium nitrate.

Example 3.9.1

$$\% \text{ composition by mass of N} = \frac{\text{mass of N in formula}}{\text{formula mass of NH}_4\text{NO}_3} \times 100$$

$$= \frac{(2 \times 14)}{(2 \times 14) + (4 \times 1) + (3 \times 16)} \times 100$$

$$= \frac{28}{80} \times 100$$

$$= 35\%$$

% composition by mass of N = 35%

Activity 3.9.4: Group activity

This activity will give you practice in calculating percentage composition.

Work in a group to calculate the percentage by mass of the essential elements in the rest of the compounds in Table 3.9.2. Each member of the group could in turn take a compound and work through the percentage calculations with the other members of the group. Check your answers with the figures in Table 3.9.2.

SPOTLIGHT ON INDUSTRY: INDUSTRIAL MANUFACTURE OF AMMONIUM NITRATE

CSBP Limited based in Kwinana, Western Australia, is an example of a company that makes ammonia, nitric acid and ammonium nitrate on one site. The flow diagrams summarise the processes which take place on an industrial scale.

The basic ammonia production process uses natural gas, steam and air to produce the ammonia.

- *First, sulfur is removed from the natural gas. Sulfur is an impurity which deactivates (poisons) the catalyst.*

- *Natural gas and steam are then reacted at approximately 1000°C to produce carbon monoxide and hydrogen in a process known as primary reforming.*

- *The carbon monoxide and hydrogen are mixed with air to produce more carbon monoxide and hydrogen gas. The air also provides nitrogen for the subsequent synthesis of ammonia. This process is known as secondary reforming.*

- *The carbon monoxide is then converted to carbon dioxide and removed.*

- *Any remaining carbon oxides are converted to methane in a process known as methanation.*

- *Hydrogen and nitrogen are reacted over a catalyst (the Haber process) to produce ammonia.*

- *The ammonia gas is refrigerated and converted to liquid for storage.*

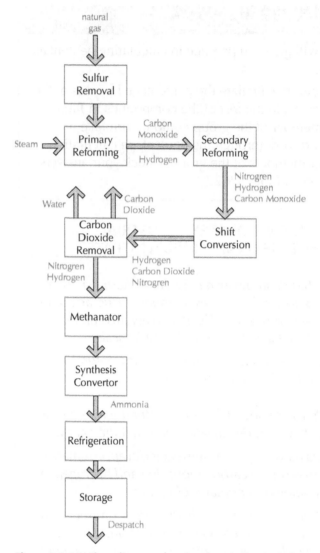

Figure 3.9.10: *Flow diagram showing the production of ammonia at the CSBP plant in Western Australia*

The ammonia plant is capable of producing up to 740 tonnes of ammonia a day. The plant has an environmentally-efficient design and refrigerated storage system .

The carbon dioxide produced as a by-product is used as a water treatment chemical in the food and beverage industry and in medical applications.

CSBP operate two modern nitric acid/ammonium nitrate production plants and a prill facility at Kwinana. Prilling is the process where concentrated ammonium nitrate is sprayed into a tower, the water evaporates and as the concentrated droplets fall they join together and form solid spheres, known as prills.

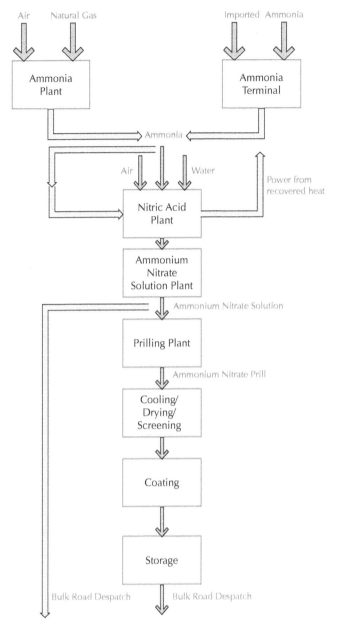

Figure 3.9.11: *Flow diagram showing the industrial manufacture of ammonium nitrate at the CSBP plant in Western Australia*

Production process:

- *First, ammonia is reacted with oxygen from the air and absorbed into water to form nitric acid (the Ostwald process).*
- *The nitric acid is then combined with ammonia in a reactor to produce ammonium nitrate solution.*
- *The ammonium nitrate solution is concentrated and sprayed into a prilling tower to produce dry prills.*
- *The prills of ammonium nitrate can be coated with chemicals such as fungicides to help the plants fight disease.*

Most of the nitric acid is reacted with ammonia to make ammonium nitrate and a small amount is sold for other uses. Some of the ammonium nitrate produced is used in the fertiliser industry but most goes into making explosives to support the ever increasing demand from Australia's expanding mining industries.

Environment watch

Eutrophication

Overuse of nitrate and phosphate fertilisers can result in nitrate and phosphate ions being washed into rivers and lakes. These ions are nutrients for all plant life in the water and feed the green algae which are in most rivers. This results in an algal bloom forming on the surface of the water. The algae prevent sunlight from reaching plants that grow beneath the surface of the water. This in turn means the plants can't photosynthesise and they die. Bacteria feed on the dead plants and use up oxygen in the water. As a result other forms of life such as fish die. The whole process – which results in the death of a river or lake – is known as eutrophication.

Figure 3.9.12: *A river covered with algal bloom due to nitrate and phosphate ions being washed from farmland*

◗◖ SPOTLIGHT ON FOOD PRODUCTION IN THE FUTURE

The world's population continues to grow and the demand for food grows with it. There is a limit to the amount of land available for growing food so we have to be smarter about how we use this land, as well as finding ways of using both land which at present can't be used and marine environments.

Milk drinks that lower blood pressure, meat products that reduce the risk of heart disease, chocolate that calms you down and a new range of foods that can fight obesity can be created from marine animals and plants. Japan already has several product ranges on the shelves and research programmes are in place all over the world to create more.

Figure 3.9.13: *Barley growing in the desert of Qatar*

A desert is the last place you might think of for growing food but this is what is planned for the desert of Qatar. It is a very tough part of the world to grow food. Fresh water is very scarce, a lot of it obtained by the desalination (removing the salt) of seawater. In February 2012 an agreement was signed between the fertiliser company Yara International ASA, the Qatar Fertiliser company QAFCO and The Sahara Forest

🔍 Hint

This is only the start – you can look for updates over the coming years by entering the Sahara Forest Project into an internet search engine.

Project AS to cooperate on a pilot project in Doha, Qatar. This pioneering environmental project will use seawater, greenhouses, solar energy, carbon dioxide and fertilisers for cultivating desert land and making it green. The aim is to do this sustainably and at an acceptable cost. In the greenhouses seawater will be used to provide cool and humid growing conditions for fresh vegetables. The greenhouse will also produce fresh water. The experiment will include use of concentrated solar power for generating clean energy and growing plants which are tolerant of salty water.

Learning checklist

In this chapter you have learned:

- Growing plants need nutrients, including compounds of nitrogen, phosphores and potassium.
- Fertilisers are substances which return nutrients to the soil.
- Ammonia is an important feedstock for the manufacture of fertilisers.
- Ammonia can be made in the laboratory by heating an ammonium salt with a base.
- Ammonia is a very pungent soluble gas which forms an alkaline solution when it dissolves in water.
- Ammonia is made industrially by the Haber process.
- In the Haber process nitrogen from the air combines with hydrogen from the petrochemical industry.
- The reaction of nitrogen with hydrogen to make ammonia is reversible and the ammonia breaks down if the temperature is too high.
- The Haber process needs a temperature of 450°C, pressure of 200 atmospheres and an iron catalyst.
- Nitric acid is an important feedstock for the manufacture of ammonium nitrate.
- Nitric acid is made industrially by the Ostwald process.
- In the Ostwald process ammonia and oxygen are passed over a platinum catalyst at 900°C.

- The reaction is exothermic so external heat can be removed when the reaction gets started.

- Ammonia reacts with acids to form soluble salts.

- Nitric acid is used to make ammonium nitrate.

- Ammonium nitrate is formed when ammonia and nitric acid react in a neutralisation reaction.

- Ammonium nitrate is used as a fertiliser.

- How to calculate the percentage composition of nitrogen in ammonium nitrate and other fertilisers.

10 Nuclear chemistry

You should already know

- Atomic structure in terms of protons, neutrons and electrons.
- Isotopes as atoms of the same element having different mass numbers.
- Nuclide notation as a means of representing individual isotopes.

Learning intentions

In this chapter you will learn about:

- Radiation.
- Use of radioactive isotopes.
- Nuclear equations.
- Half-life.

⌕ Hint

Radiation is also known as radioactive decay and radioactivity.

Radiation

The atoms of most elements have isotopes. The nuclei of some of these isotopes are unstable and give out particles and rays, called emissions. This is known as radioactivity. Radioactivity happens spontaneously, no matter what state the element is in or if it is chemically combined in a compound. These emissions are called alpha (α), beta (β) and gamma (γ). Radioactivity is all around us – this is known as background radiation.

Figures 3.10.1 and 3.10.2 show the penetrating power of the radioactive emissions and what happens to them when they are passed through an electric field.

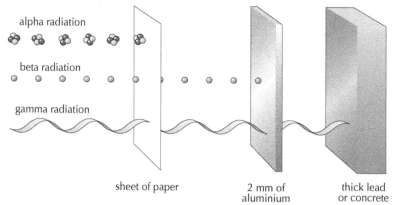

alpha radiation

beta radiation

gamma radiation

sheet of paper

2 mm of aluminium

thick lead or concrete

Figure 3.10.1: *Penetrating power of radioactive emissions*

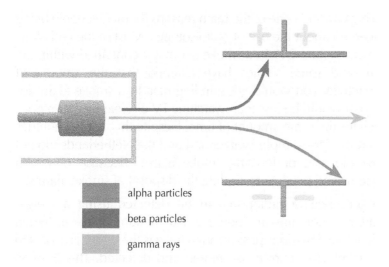

Figure 3.10.2: *Deflection of emissions in an electric field*

α particles are the heaviest of the emissions and are made up of two protons and two neutrons grouped together. This results in them having a positive charge overall. An α particle is in fact a helium ion – He^{2+}.

Its nuclide notation is $^{4}_{2}He^{2+}$ – the charge is often not shown.

β particles are high energy electrons which can travel a few metres in the air. At first glance this should seem unlikely because electrons are not found in the nucleus. They are formed when a neutron in the nucleus breaks down and a proton and velectron are formed. The proton stays in the nucleus and the electron is emitted. A β particle is often represented as $^{0}_{-1}e$.

γ radiation is not a particle and so has no mass and no charge. It is a high energy ray which can travel long distances and has high penetration power.

GO! **Activity 3.10.1**

Use all the information to complete the table.

Name of emission	Distance travelled and penetrating power	Charge
(a)	Few cm in air. Stopped by paper.	(b)
β	(c)	(d)
(e)	(f)	None

If radioactive emissions enter the body of animals they can knock electrons out of molecules which make up living tissue. This results in the formation of positively charged ions which can cause biological damage. The fear of damaging healthy cells and causing cancer is one of the main reasons for many people being concerned about the use of radioisotopes. Most of the emissions from americium-241 in a smoke alarm are contained within the alarm and some studies have claimed that the amount of radioactivity you would get standing next to a smoke alarm for one year would be less than watching TV or wearing a luminous watch for the same amount of time. Despite this, some countries in Europe, for example Switzerland and the Netherlands, do not allow the sale of ionising smoke alarms because of health concerns and issues surrounding the disposal of smoke alarms.

The presence of radiation can be detected using a Geiger counter. It contains an inert gas like helium or neon. When radioactive emissions pass through the gas the atoms are ionised and an electric current is created and detected. This is often converted into the well-known clicking sound heard coming from the Geiger counter.

STEP BACK IN TIME

The term radioactivity was first used by Marie and Pierre Curie over 100 years ago. They worked together to show that radioactivity is an atomic property and not a chemical change. They were not the first to observe the effects of radioactivity. In 1896 a French scientist called Henri Becquerel noticed that a photographic plate left in the dark next to a uranium salt became fogged as if it had been exposed to light. The fogging of the plate was due to radioactive emissions from the uranium.

The work of Becquerel and the Curies was so important that they were awarded the Nobel prize for physics in 1903. Marie Curie was awarded a second Nobel prize in 1911, this time for chemistry, for isolating the radioactive element radium. In that same year the Sorbonne University in Paris built the first radium institute with two laboratories – one for the study of radiation and the other for biological research into treatment of cancer. Marie Curie died from leukaemia in 1934. It is thought that exposure to radioactivity over many years caused this form of cancer. In 1948 the Marie Curie Memorial Foundation was established and it became a charity dedicated to relieving suffering from cancer by establishing specialist homes for the care of cancer patients and providing nursing for patients in their own home. The Foundation led the way in educating the public about the symptoms and treatment of cancer and provided urgent welfare needs for patients and their families. Today the foundation is known as Marie Curie Cancer Care and still provides services for cancer patients and also carries out research to find out what the best possible care is and how to provide it.

Figure 3.10.3: *Marie Curie in her laboratory*

GO! Activity 3.10.2

Find out more about the work of Marie Curie Cancer Care. Their website is a good source of information.

Figure 3.10.4: *The Marie Curie Cancer Care logo*

Figure 3.10.5: *Cover from an ionising smoke alarm showing it uses americium-241*

Use of radioactive isotopes

'Everyday' uses

Radioactive isotopes are also known as radioisotopes and they are important in many everyday situations. In the home ionising smoke alarms contain very small amounts of americium-241 (as americium dioxide) which emit α particles which ionise the gases in air and cause an electric current. When smoke passes into the alarm's detector the current drops and the alarm sounds. Only a few kilograms of americium-241 are made each year and most is used in ionising smoke detectors for the home. It is estimated that one gram of americium-241 is enough to make more than three million smoke alarms.

In airports, luggage or clothing suspected of being contaminated with explosives is swabbed with a piece of cloth and put into a detector containing nickel-63 which emits β radiation.

Some radioisotopes are used to irradiate food – this means the food is deliberately exposed to radiation. Cobalt-60 emits high energy γ rays which kill bacteria such as salmonella and E-coli which cause food poisoning. The effect is similar to cooking or pasteurising but irradiating does not change the appearance or texture of the food too much. Irradiation of food is hardly used in the UK while it is widely used in countries like Brazil.

The Radura logo is the international symbol to indicate that food has been irradiated to kill bacteria.

Medical uses

The health service uses radioisotopes in a variety of ways. Cancer in the thyroid glands in the neck can be detected and treated using isotopes of iodine which emit γ and β radiation. Technitium-99m is the most widely used radioisotope for diagnostic studies in nuclear medicine. Different chemical forms are used for brain, bone, liver, spleen and kidney imaging and also for blood flow studies.

Figure 3.10.6: *Explosive trace detection (ETD)*

Figure 3.10.7: *Irradiated food, showing the Radura logo*

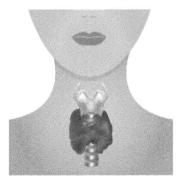

Figure 3.10.8: *Radioactive isotope of iodine are used to detect and treat cancer in the thyroid glands in the throat*

Industrial uses

Adhesive tape, cling film, paper, vinyl floor coverings, plastic sheeting and the sheets of steel used to make car bodywork: what do they all have in common? They are all products which rely on the use of radioisotopes to help control their thickness – this is known as gauging. Gauging uses the penetrating power of radioactive emissions.

The material passes between a radioactive source and a detector. The intensity of the radioactivity reduces when it hits the material. This is picked up by the detector and the thickness of the material can be adjusted as necessary.

The same idea is used to measure the level of the liquid in a can of soft drink.

Caesium-137 and cobalt-60 are radioisotopes commonly used in gauging.

Figure 3.10.9: *The thickness of sheet metal can be controlled using radioisotopes*

Radioisotopes are used as tracers in industry. They are used to find out what is happening inside objects without the need to break into it.

To locate a leak in an underground pipe, for example, a very small amount of radioactive material that gives off gamma rays such as sodium-24 is put into the pipe. A detector is moved along the ground above the pipe. The leak is located where there's an increase in activity and little or no activity after that point.

Iodine-131 is used as a tracer in an industrial process called hydraulic fracking which extracts hydrocarbons trapped in shale deposits underground – hydraulic fracking is covered in more detail in Area 2, Chapter 5.

GO! Activity 3.10.3: Paired activity

1. Work on this activity with a partner. You may wish to share the research but present your findings individually.

   - Look at the uses of radioisotopes in industry and medicine described in the text above.

   - Select one of the radioisotopes and find out the detail of how it is used and how it is of benefit to us.

   - Present your findings. You can do this any way you want but here are some suggestions:

     » Write a short report (50–100 words).

     » Make up a slide presentation using, for example, Power Point.

     » Make a poster which can be displayed on the wall.

     » Put together a commentary you could use to present to the rest of the class.

     You may wish to access a website search engine and find out more about the uses of radioisotopes in industry and medicine – you could include other uses if you wish.

     Use key words such as 'radioisotopes' or some of the other words and phrases in the text.

2. Complete the summary by filling in the missing words. You can use the word bank to help you. You may wish to copy the completed summary into your notes.

   Atoms with unstable __(a)__ can emit __(b)__ (α), beta (β) and __(c)__ (γ) radiation. α particles are __(d)__ nuclei and β particles are high energy __(e)__. α particles are __(f)__ moving and can be stopped by paper. __(g)__ particles are fast moving and can be stopped by a thin sheet of __(h)__. γ rays can travel long distances and can only be stopped by thick lead or concrete. Radioisotopes have many everyday uses. They are used in __(i)__ alarms, in the detection of explosives and in the detection and treatment of __(j)__. Radiation __(k)__ molecules as it passes through them which causes damage to human tissue. In industry, radiation can be used to gauge the __(l)__ of materials and as __(m)__.

   Word bank: alpha, aluminium, beta, cancer, electrons, gamma, helium, ionises, nuclei, slow, smoke, thickness, tracers

Nuclear equations

The breakdown of unstable nuclei (radiation) is also known as decay. An atom can decay through a series of stages until it becomes stable.

The emission of an α particle means the nucleus of the original atom loses two protons and two neutrons. The emission of a β particle results in the original atom losing a neutron but gaining a proton. γ radiation has no effect on the original atom because it is not a particle.

Radioactive decay can be represented by nuclear equations – they can be used to summarise the processes which produce α and β radiation.

Nuclear equations include the mass number (number of protons + neutrons), the atomic number (number of protons) and the chemical symbol for each particle involved, i.e. nuclide notation.

Example 3.10.1

When the americium-241 used in smoke detectors emits α particles (helium nuclei) each atom loses two protons and two neutrons. This results in the americium-241 atom changing into a neptunium-237 atom:

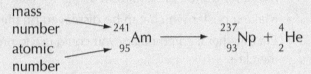

$$\text{mass number} \nearrow \quad \text{atomic number} \nearrow \quad {}^{241}_{95}\text{Am} \longrightarrow {}^{237}_{93}\text{Np} + {}^{4}_{2}\text{He}$$

Note that the total of the mass number on the left of the arrow must equal the total on the right of the arrow. It is the same for the atomic numbers.

Example 3.10.2

When the nickel-63 used to detect explosives emits β particles (electrons) each atom gains a proton (remember a proton and electron are formed from a neutron). This results in the nickel-63 atom changing into a copper-63 atom.

$$^{63}_{28}\text{Ni} \longrightarrow {}^{63}_{29}\text{Cu} + {}^{0}_{-1}\text{e}$$

Note again that when the mass numbers and atomic numbers are added on, each side of the arrow is the same. The electron (β particle) is given an unusual atomic number (-1). This is a way of indicating that an extra proton is gained by the parent atom when a beta particle is emitted and so the rule that the total atomic number must be the same on each side of the arrow is satisfied.

So long as you know the particle being emitted the element formed can be worked out as illustrated in Example 3.10.3.

Example 3.10.3

Work out what X, a and b are in the nuclear equation:

$$^{220}_{86}\text{Rn} \longrightarrow {}^{a}_{b}\text{X} + {}^{4}_{2}\text{He}$$

(Answer: Applying the rules, a = 216, b = 84 so X = Po, i.e. ${}^{216}_{84}\text{Po}$)

Activity 3.10.4

1. Work out what X, a and b are in the nuclear equation:

$$^{216}_{84}\text{Po} \longrightarrow ^{a}_{b}\text{X} + ^{0}_{-1}\text{e}$$

2. Write the equation for the α decay of $^{234}_{92}\text{U}$.

3. Write the equation for the β decay of $^{228}_{89}\text{Ac}$.

Artificial radioisotopes can be made in nuclear reactors by bombarding stable nuclei with neutrons. For example:

$$^{27}_{13}\text{Al} + ^{1}_{0}\text{n} \longrightarrow ^{24}_{11}\text{Na} + ^{4}_{2}\text{He}$$

The sodium isotope produced is radioactive and decays itself by β emission:

$$^{24}_{11}\text{Na} \longrightarrow ^{24}_{12}\text{Mg} + ^{0}_{-1}\text{e}$$

Many artificial radioisotopes are produced for specific uses in healthcare and industry.

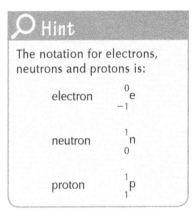

Hint

The notation for electrons, neutrons and protons is:

electron	$^{0}_{-1}\text{e}$
neutron	$^{1}_{0}\text{n}$
proton	$^{1}_{1}\text{p}$

Half-life

The nuclei of radioisotopes decay in a random fashion. The time in which half of the nuclei of a radioisotope would be expected to decay is known as the half-life. Each radioisotope has a unique half-life and this can vary from fractions of a second to billions of years.

Table 3.10.1: *Table of radioisotope half-lives*

Radioisotope	Half-life	Use
chromium-51	27·7 days	Labelling of red blood cells
iodine-131	8·02 days	Diagnosing/treating diseases associated with the thyroid gland
phosphorus-32	14·28 days	Treatment of excess red blood cells
technetium-99m	6·01 hours	Imaging the organs of the body
americium-241	433 years	Neutron gauging and smoke detectors
cobalt-60	5·27 years	Gamma radiography and medical equipment sterilisation
caesium-137	30·07 years	Thickness gauging
gold-198	2·7 days	Tracing factory waste causing pollution

As the atoms of a radioisotope decay, the intensity of the radiation decreases. After one half-life the intensity of the radiations will have fallen to half its original value. After a second half-life the intensity of the radiation will have halved again, i.e. it will be one-quarter of its original value. A graph of the intensity of the radiation against time gives a radioactive decay curve with a typical shape.

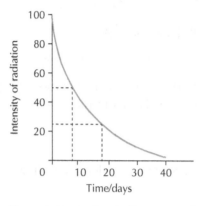

Figure 3.10.10: *Graph of intensity of radiation against time for a sample of iodine-131*

Figure 3.10.11 shows it takes eight days for the intensity of the radiation to halve, i.e. the half-life for iodine-131 is eight days.

In two half-lives the intensity of the radiation will have dropped to one quarter of its original value.

Radioisotopes with a short half-life are used in diagnosing and treating medical conditions. This is because radiation damages healthy cells, so the shorter the length of time the radioisotope is active, the less damage is done to healthy cells.

GO! Activity 3.10.5

1. An isotope of technicium (^{99}Tc-*m*) used in medical imaging decays by gamma emission. The half-life for the transition is six hours. Draw the decay curve showing how the activity of a sample would decrease over a period of two days. Take the initial activity to be 100%.

2. The half-life of americium-241 used in smoke alarms is nearly 433 years. Suggest why its half-life makes it useful for use in household smoke alarms.

Environment watch

Dalgety Bay, Fife, Scotland

The following article appeared on the Scottish Environment Protection Agency (SEPA) website in August 2012:

'Radioactive material was first detected on a part of the foreshore at Dalgety Bay in 1990. Monitoring has been undertaken by both SEPA and the Ministry of Defence and periodically

Hint

You can keep up-to-date with what's happening about radioactivity in Dalgety Bay by checking the SEPA website.

radioactive material has been removed. It is thought that the contamination originates from the residue of radium-coated instrument panels from military aircraft incinerated and land-filled in the area at end of World War II.

The radium used by the MoD was primarily in luminescent paints. Radium-based luminescent paint was typically made by mixing a radium salt, zinc sulphide and a carrier material (typically varnish or lacquer).'

> ## ⚠ Think about and discuss
>
> Why do you think the residents of Dalgety Bay are concerned about the discovery of radioactive material in the bay?

Radiocarbon dating

The element carbon has three isotopes. Two of the isotopes, carbon-12 and carbon-13, are stable. The third, carbon-14, is unstable.

Atoms of carbon-14 are formed in the upper atmosphere when neutrons in cosmic rays collide with nitrogen atoms.

$$^{14}_{7}N + ^{1}_{0}n \longrightarrow ^{14}_{6}C + ^{1}_{1}p$$

The carbon atoms are quickly oxidised to carbon dioxide, which is absorbed by plants during photosynthesis and passed into food chains. All living things therefore contain a small amount of carbon-14. The proportion of carbon-14 in a living organism remains constant throughout its lifetime.

Carbon-14 decays by beta emission. It has a half-life of approximately 5730 years.

$$^{14}_{6}C \longrightarrow ^{14}_{7}N + ^{0}_{-1}e$$

When an organism dies the amount of carbon-14 in the organism begins to decrease as carbon-14 atoms decay. Measuring the radioactivity of a sample of carbon from an article and comparing it to the activity of a modern day sample allows an article to be dated. This method has been used extensively to date archeological artefacts from areas such as ancient Egypt.

Figure 3.10.11: *Ötzi – the Iceman: carbon-14 dating shows that Ötzi lived about 3300 BCE*

Dating rocks and fossils

Radiocarbon dating is only useful for once living materials less than about 50 000 years old due to the very small amount of carbon-14 present in materials older than this.

Figure 3.10.12: *Igneous rocks are formed when molten magma solidifies*

Other methods making use of radioisotopes with much longer half-lives have to be used to date rocks and fossils.

Volcanic and igneous rocks can be dated using an isotope of potassium. Potassium-40 decays to form argon-40. The half-life for this transition is 1·3 billion years.

After the rocks solidify any argon-40 produced by the decay of potassium-40 will be trapped in the rock. The proportion of potassium-40 to argon-40 in a rock sample can then be used to date the age of the rock.

GO! Activity 3.10.6: Paired activity

The following equation describes the transition when potassium-40 decays to argon-40.

$$^{40}_{19}\text{K} + ^{0}_{-1}\text{e} \longrightarrow ^{40}_{18}\text{Ar} + \gamma$$

Discuss with a partner the changes that take place.

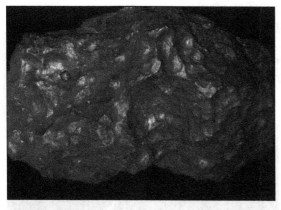

Figure 3.10.13: *A fragment of the Canyon Diablo meteorite*

GO! Activity 3.10.7

Write a nuclear equation which represents the decay of Rb-87 to Sr-87. Use the SQA data booklet to find the atomic numbers.

Dating the Earth

Scientists are unable to deduce the age of the Earth from rocks because Earth rocks have been formed and re-formed over time. They have deduced the age of the Earth by dating meteorites which they believed were formed at the same time as the Earth. Isotopes of uranium in meteorites decay to form isotopes of lead. Measuring the proportions of lead isotopes to uranium isotopes gives an indication of the age of the meteorites.

This can also be done by determining the proportion of Sr-87 to Rb-87 in minerals in the meteorites.

The largest meteorite ever found weighed 60 tonnes and was found in Namibia but larger meteorites have struck the Earth. The Canyon Diablo meteorite which caused the formation of Meteor Crater in Arizona when it struck the Earth is estimated to have weighed over 60 000 tonnes. Analysis of meteorites indicates that the Earth is about 4·5 billion years old.

Learning Checklist

In this chapter you have learned:

- There are many unstable isotopes of elements.
- Unstable isotopes can become more stable by emitting radiations.
- Isotopes which emit radiations are known as radioactive isotopes or radioisotopes.
- Radioisotopes have many important industrial and medical uses.
- The three types of radiation emitted from nuclei are alpha (α), beta (β) and gamma (γ).
- Alpha and beta radiations change an isotope of one element to an isotope of another element.
- Gamma emissions are due to nuclei losing energy.
- Alpha particles are helium nuclei ($^{4}_{2}He^{2+}$), i.e. they are heavy, positively charged particles.
- Alpha particles are slow moving and have low penetration. They will only travel a few centimetres through air and are stopped by paper.
- Beta particles are electrons ($^{0}_{-1}e$) and are fast moving, i.e. they are negatively charged.
- Beta particles are more penetrating than alpha particles but are stopped by a thin sheet of aluminium.
- Nuclear equations are used to describe the transitions which produce radiations.
- The mass numbers and atomic numbers of isotopes are shown in nuclear equations.
- The time in which half of the nuclei of a radioisotope decay is known as the half-life.
- Half-lives for radioisotopes are unique and constant.
- The age of materials can be dated using the half-lives of radioisotopes.

11 Chemical analysis

You should already know:

- Chromatography is a technique which can be used to separate substances, e.g. the colours that make up the ink in a pen.
- pH measurement can be carried out using pH meters or indicators where different colours indicate different pH values.
- Separation techniques include dissolving and filtering. For example, a salt and sand mixture can be separated by adding the mixture to water – the salt will dissolve allowing the sand and salt to be separated by filtering.

Learning intentions

In this chapter you will learn about:

- Common chemical apparatus.
- General practical techniques.
- Qualitative analysis: indicating the substances present.
- Quantitative analysis: determining how much of a substance is present.
- Reporting experimental work.

Figure 3.11.1: *A tractor spraying crops with fertiliser*

The role of chemists

Chemists have an important role in monitoring our environment. They monitor air and water quality as well as monitoring soil quality to ensure that farmers can grow crops under optimum conditions.

Pollution from fertiliser run-off from fields has caused algal blooms in rivers and lochs leading to the death of aquatic plants and fish.

⊛ STEP BACK IN TIME

In the past the main source of air pollution was from the burning of fossil fuels both in homes and by industry. The city of Edinburgh is known as Auld Reekie, which means Old Smoky. It was given this name because of the smoke given off from the tenements and close-packed housing of the city.

In December 1952, London was affected by smog. Smog is a mixture of smoke and fog. It was so thick that vehicles, including public transport vehicles, were abandoned. This brought the city to a halt. A bonus for schoolchildren was that schools had to be closed.

Although it lasted only four days, it is estimated that over 4000 people, and perhaps as many as 12 000, died prematurely from respiratory conditions brought on by the smog.

Figure 3.11.3: *The reddish-pink sandstone of the Scott Monument in Edinburgh, like many of the city's buildings, has been blackened by the smoke from domestic fires*

Figure 3.11.2: *The smog was the result of smoke from coal burning mixing with fog*

The Great Smog of 1952 led to Parliament passing the Clean Air Act. The act led to the use of smokeless fuels replacing coal for household fires in cities.

Not only did the burning of coal produce smoke pollution and increase carbon dioxide levels in the atmosphere but because fossil fuels such as coal contain sulfur compounds, sulfur dioxide was emitted into the atmosphere leading to acid rain.

▄▄◀ SPOTLIGHT ON POLLUTION FROM TRANSPORT

Changes in the law have led to more efficient burning of fossil fuels and the removal of harmful gases before emissions are released into the atmosphere.

Air pollution from domestic and industrial sources is tending to decrease. However air pollution caused by transport is increasing worldwide. The main air pollution problem nowadays comes from the increasing number of vehicles on our roads. Burning petrol, diesel and kerosene emits a variety of pollutants into the atmosphere. These include carbon monoxide (CO), oxides of nitrogen (NO$_x$), volatile organic compounds (VOCs) and particulate matter (PM$_{10}$). Particulate matter (PM$_{10}$) consists of particles smaller than 10 micrometres (1 × 10^{-5}m). These can pass through the nose and throat and enter the lungs.

Air pollution in our towns and cities is a particular problem.

Figure 3.11.4: *Many cyclists in towns and cities wear protective masks or scarves to try to prevent breathing in particulate matter*

Adelaide in Australia has good air quality with very low air pollution levels. But even there 1178 tonnes of sulfur dioxide were still released into the atmosphere in 2002. Table 3.11.1 shows the different sources of sulfur dioxide that affected the city.

Table 3.11.1: *Sources of sulfur dioxide for Adelaide in 2002*

Source	Estimated emissions (tonnes/year)	% of total
Motor vehicles	666·7	(a)
Gaseous fuel burning (domestic)	1·7	0·14
Liquid fuel burning (domestic)	8·0	0·70
Solid fuel burning (domestic)	45·7	3·90
Lawn mowing	(b)	4·23
Commercial shipping	216·2	(c)
Recreational boating	0·2	0·02
Aeroplanes	39·2	3·33
Railways	10·1	0·90
Other fuel combustion	(d)	12·00
Total emissions	**1178**	**100**

Source: Environment Protection Authority South Australia

⊙ Activity 3.11.1

Calculate values for the missing entries in Table 3.11.1.

◄€ SPOTLIGHT ON THE ENVIRONMENT

Pollution levels in the environment need to be monitored because of the potential impacts on human health and on the natural environment itself.

Satellites launched into orbit above the Earth can give details of atmospheric pollution. For example, from its launch in 2002 until 2012 when contact with the satellite was lost, instruments on the Envisat satellite recorded the spectrum of sunlight shining through the atmosphere and determined the concentration of pollutant gases in the atmosphere. The European Space Agency plans to launch six Sentinel satellites between 2013 and 2020 to continue monitoring conditions on Earth and in the atmosphere.

Figure 3.11.5: *Envisat, Earth observation satellite. The satellite was launched into polar orbit 790 km above the Earth on 1 March 2002 by the European Space Agency*

Figure 3.11.6: *Nitrogen dioxide pollution map. Satellite image-based map, showing the average nitrogen dioxide (NO2) produced around the world in 2006, from blue (lowest levels) to red (highest levels).*

Figure 3.11.7: *An automatic air pollution monitoring station. Readings are taken every hour*

Concentrations for a number of air pollutants are monitored continuously at a wide range of urban and rural sites throughout the UK. In London air quality is monitored at over 100 sites. Stations on the Automatic Urban and Rural Network make measurements of pollutants such as nitrogen dioxide every hour.

GO! Activity 3.11.2: Paired activity

The graph shows how nitrogen dioxide levels in Inverness changed during a week in 2012.

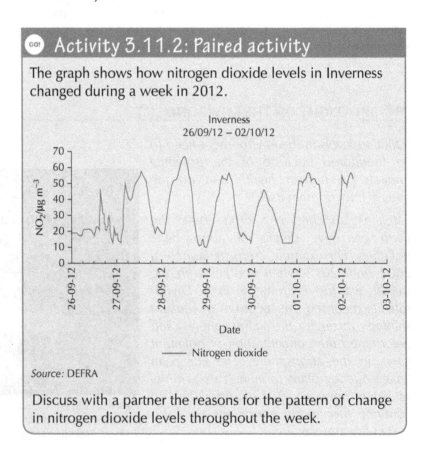

Source: DEFRA

Discuss with a partner the reasons for the pattern of change in nitrogen dioxide levels throughout the week.

The economic crisis has had an effect on air pollution. Following the downturn in 2008 levels of oxides of nitrogen in the air have fallen across Europe due to lower car usage and industrial activity. In 2011 levels of sulfur dioxide and oxides of nitrogen in Greece had decreased by as much as 40% from 2008 values. However any health benefits were offset by smoke pollution caused by wood burning. In 2012 a new tax pushed up the price of heating oil by a third. Many people, unable to afford heating oil, turned to wood to heat their homes. Mount Olympus is the site of some of the worst illegal logging in Greece. The practice has risen here by more than 300%.

SPOTLIGHT ON INDUSTRY – PREVENTING AIR POLLUTION

Chemists and chemical engineers have an important role in devising ways of removing harmful chemicals and thus preventing them affecting the environment.

Many industrial plants such as coal burning power stations remove sulfur dioxide by a process known as flue gas desulfurisation (FGD). **Flue gases** are passed through a scrubber tower before being released into the atmosphere.

Sulfur dioxide is an acidic gas and is very soluble. If flue gases are passed through an alkaline spray the sulfur dioxide will react. Some wet scrubbers use a limestone slurry (finely powdered limestone suspended in water). The sulfur dioxide is converted to calcium sulfite.

$$CaCO_3(s) + SO_2(g) \rightarrow CaSO_3(s) + CO_2(g)$$

The calcium sulfite can be oxidised to calcium sulfate by blowing compressed air through it. Some of the costs of FGD can be recouped by selling the calcium sulfate, which is used to make plasterboard.

The carbon dioxide produced cannot be allowed to escape into the atmosphere as it is thought to contribute to global warming. Carbon dioxide is an acidic gas so can be scrubbed with alkalis such as sodium hydroxide and calcium oxide (quicklime) in a neutralisation reaction.

📖 Word bank

• **Flue gas**

Gases such as sulfur dioxide and carbon dioxide commonly produced as by-products of industrial processes.

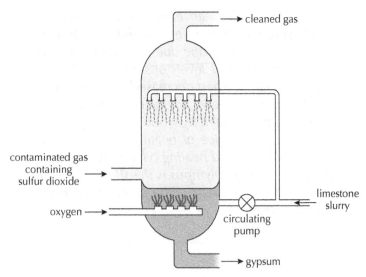

Figure 3.11.8: *Limestone slurry sprayed in at the top of the scrubbing tower removes harmful sulfur dioxide from the emissions*

It is not only in industry that emissions are being controlled. In the United States controls are being tightened on emissions from small petrol engines used for lawnmowers, chainsaws and leaf blowers. Legislation has meant emission of unburned hydrocarbons has had to be reduced by nearly 80%. Manufacturers have developed catalysts which help achieve the new limits for small engines.

📖 Word bank

- **Photocatalysts**
Catalysts that absorb light energy.

🔦 SPOTLIGHT ON TECHNOLOGY – PHOTOCATALYSTS

A novel approach to dealing with urban traffic-related pollution is the suggestion that **photocatalysts** *could be incorporated into clothing. The catalysts absorb light energy and would help in the breakdown of pollutants to harmless products. The photocatalyst could be added to clothes as part of the fabric conditioner in a normal washing cycle.*

Figure 3.11.9: *This dress is impregnated with a photocatalyst that uses light to break down air-borne pollution into harmless chemicals*

Photocatalysts are already incorporated into coatings for concrete structures and coatings for self-cleaning glass. It is estimated that if 15% of a city's buildings were coated with photocatalytic products then air pollution might be reduced by as much as 50%.

Self-cleaning windows use a thin coating of titanium dioxide which acts as a photocatalyst. In the presence of light the photocatalyst breaks down pollutants including organic materials that stick to the window surface. The titanium dioxide surface is hydrophilic (water loving). Rainwater falling on the window forms a thin film rather than droplets. The thin film washes off any materials that have been formed by the breakdown of pollutants cleaning the windows.

Figure 3.11.10: *The refurbished roof of King's Cross station was constructed using 1200 tonnes of steel and self-cleaning glass*

In Japan, scientists have developed self-cleaning windows for the Bullet train using a thin coating of the metal niobium as photocatalyst.

Figure 3.11.11: *Shinkansen Bullet train*

Common chemical apparatus

You should be familiar with and know how and when to use the pieces of apparatus listed below.

- Beaker
- Burette
- Conical flask
- Delivery tubes
- Dropper
- Evaporating basin
- Filter funnel
- Measuring cylinder
- Pipette and safety filler
- Test tube/boiling tube
- Thermometer

General practical techniques

You need to be familiar with the following techniques and be able to draw labelled, sectional diagrams for common apparatus.

Filtration

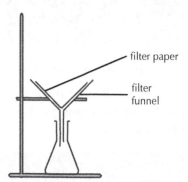

filter paper

filter funnel

Figure 3.11.12: *Filtration set up in the laboratory and labelled sectional diagram*

Figure 3.11.13: *Measuring the mass of a substance using a top pan balance*

Using a balance

A balance allows you to accurately weigh the mass of a substance. Typical school balances will measure masses in grams correct to one or two decimal places. More accurate balances used for analytical work will measure to four decimal places.

Collecting gases

The sectional diagrams show common methods of gas collection. The method depends on the solubility of the gas and how pure a sample is required.

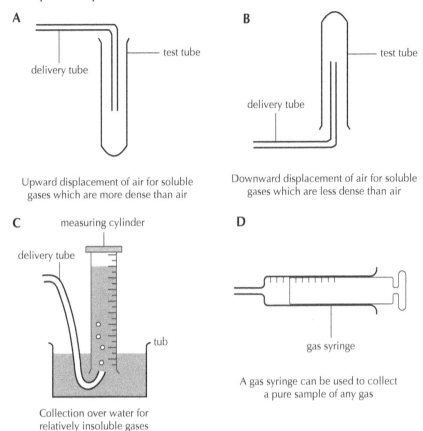

A

delivery tube — test tube

Upward displacement of air for soluble gases which are more dense than air

B

delivery tube — test tube

Downward displacement of air for soluble gases which are less dense than air

C measuring cylinder

delivery tube

tub

Collection over water for relatively insoluble gases

D

gas syringe

A gas syringe can be used to collect a pure sample of any gas

Figure 3.11.14: *Sectional diagrams showing the different methods of collecting a gas*

Heating methods

The two most common heating methods used in school laboratories are using a Bunsen burner or electric hot plate. A Bunsen burner can reach very high temperatures. For safety reasons it should not be used for heating flammable substances like alcohol. A hot plate is used when lower temperatures are required and can be used with flammable liquids like alcohol as there is no naked flame.

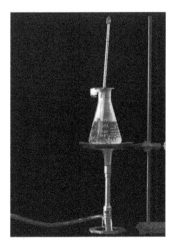

Figure 3.11.16: *A Bunsen burner can heat substances to a high temperature*

Figure 3.11.15: *A hot plate can be safely used to heat flammable liquids*

Hint

A solubility table can be found in the SQA data booklet.

Preparation of soluble salts

1. Acids with metals (see page 149).

2. Acids with bases – metal oxides, metal hydroxides and metal carbonates (see page 87).

Preparation of insoluble salts

Dissolve soluble compounds containing the ions required in the insoluble salt, in separate beakers.

Mix the solutions – a precipitate (the insoluble salt) will be produced.

Filter and wash the precipitate and allow it to dry (see page 88).

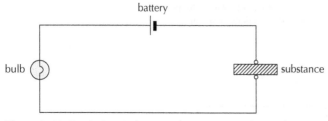

Figure 3.11.17: *A circuit diagram showing a simple conductivity tester for solid substances*

Testing electrical conductivity

A circuit diagram showing a simple conductivity tester for solid substances is shown. If the bulb lights when the substance is placed in the circuit then it is an electrical conductor. Solutions and liquids can be tested in the same way using carbon electrodes placed into the solution or liquid.

Setting up an electrochemical cell

In an electrochemical cell metal or carbon electrodes are placed in an electrolyte (ionic solution) and connected by a wire. The electrolytes and electrodes can be in separate beakers connected by an ion (salt) bridge (see page 150).

Electrolysis of solutions

Electrolysis is the breaking down of a compound by passing electricity through it. A direct current (d.c.) has to be used in order to identify the products. In a direct current one side is negative the other positive. Positive ions are attracted to the negative electrode and negative ions are attracted to the positive electrode. The diagram shows the electrolysis of copper(II) chloride solution.

Determination of E_h

The amount of energy given out (E_h) during a chemical reaction can be calculated indirectly from experimental results (see page 124).

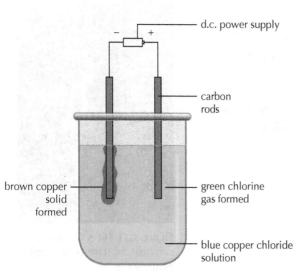

Figure 3.11.18: *The electrolysis of copper(II) chloride solution*

Qualitative analysis

Detecting substances that are present in our environment is known as qualitative analysis. The techniques used nowadays are fairly sophisticated but their origins can be traced back to simple laboratory experiments. Flame testing, precipitation reactions and chromatography are all examples of **qualitative analysis** techniques used in the laboratory.

Flame testing

When some metal compounds are placed in a flame characteristic colours are observed. For instance, when sodium compounds are placed in a flame a yellow colour is observed. Potassium gives a lilac flame. When the metal compounds are heated some electrons of the metal ions gain energy. The flame colours arise when the electrons lose the energy again. The energy is emitted as light.

Figure 3.11.19: *The lilac colour indicates that the compound in the flame contains potassium*

Carrying out a flame test in the lab

- Take a piece of Nichrome wire and form a small loop at one end.
- Clean the wire by heating the looped end then dipping it in 4 mol l^{-1} hydrochloric acid.
- Rinse the loop in deionised water.

- Dip the loop into some powdered compound or into a solution of the metal compound.
- Hold the wire to the side of the blue cone of a Bunsen flame.

Table 3.11.2: *Table of flame colours for Group 1 and Group 2 metal ions*

Group 1 Metal ion	Flame colour	Group 2 Metal ion	Flame colour
L^+	deep red	Be^{2+}	no colour
Na^+	yellow	Mg^{2+}	no colour
K^+	lilac	Ca^{2+}	brick red
Rb^+	red	Sr^{2+}	deep red
Cs^+	blue	Ba^{2+}	apple green

Figure 3.11.20: *German chemists Kirchoff (left) and Bunsen (centre) with English chemist Roscoe (right)*

🚲 **STEP BACK IN TIME**

Chemists realised that the light given off by metals when they are heated could be used to identify their presence.

The development of emission spectrometers allowed the light given off when metal compounds were heated to be examined more closely. The light given off was found to be different for each metal element and provided a signature for the element. This allowed new elements to be discovered.

Robert Bunsen and Wilhelm Kirchhoff discovered caesium, in 1860, and rubidium, in 1861, when examining spectra produced by heating lithium ores. They noticed lines on the spectra that didn't correspond to lithium. These were due to new elements. They named the elements because of particular colours they observed. Rubidium comes from the Latin word rubidius, *which means deepest red. Caesius in Latin means sky blue.*

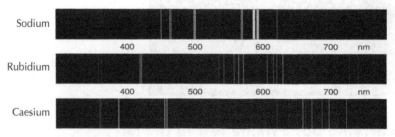

Figure 3.11.21: *The line emission spectra of sodium, and of rubidium and caesium, both discovered by Bunsen and Kirchhoff*

Precipitation

Some metal ions can also be detected using precipitation reactions.

When sodium hydroxide solution is added to solutions containing metal ions, coloured precipitates can be formed. If the solution contains cobalt(II) ions a blue precipitate will be obtained. The precipitate turns pink if more sodium hydroxide is added and the mixture heated. The precipitate is cobalt(II) hydroxide.

$$2NaOH(aq) \ + \ Co^{2+}(aq) \rightarrow \underset{pink}{Co(OH)_2(s)} \ + \ 2Na^+(aq)$$

Table 3.11.3: *Metal hydroxide precipitate colours*

Metal ion	Colour of precipitate with NaOH(aq)
Cobalt(II)	blue precipitate that turns pink if more NaOH added and mixture heated
Copper(II)	pale blue
Barium	apple green
Iron(III)	rust red
Chromium(III)	green
Silver	brown

Negative ions can also be detected using precipitation reactions.

When silver nitrate is added to solutions containing halide ions precipitates form. The colour of the precipitate indicates the particular halide ion present. When silver nitrate is added to a sodium chloride solution a white precipitate of silver chloride is obtained.

$$NaCl(aq) \ + \ AgNO_3(aq) \rightarrow NaNO_3(aq) \ + \ \underset{white}{AgCl(s)}$$

Figure 3.11.22: *Silver nitrate gives a precipitate that indicates the particular halide ion. Left: white precipitate, chloride; centre: pale cream, bromide; right: pale yellow, iodide*

GO! Activity 3.11.3: Paired activity

Discuss with a partner: if you were given a sample of calcium chloride, how would you show that it contained both calcium and chloride ions?

Testing for gases

Simple tests can be carried out to identify some common gases.

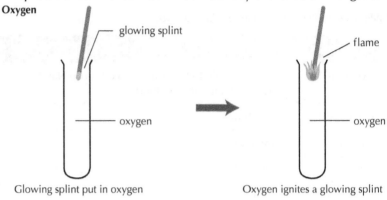

Oxygen

glowing splint

oxygen

Glowing splint put in oxygen

flame

oxygen

Oxygen ignites a glowing splint

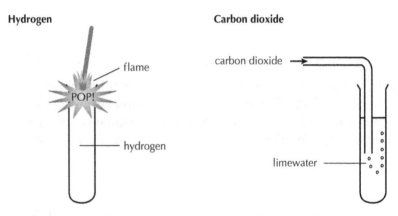

Hydrogen

flame

POP!

hydrogen

Hydrogen burns with a 'pop'

Carbon dioxide

carbon dioxide →

limewater

Carbon dioxide turns limewater cloudy

Figure 3.11.23: *Tests to identify common gases*

Methods for collecting gases can be found on page 211.

☀ SPOTLIGHT ON SPACE EXPLORATION

2012 saw the Curiosity Rover land on Mars for a mission lasting two years investigating whether Mars in the past was able to support microbial life.

Figure 3.11.24: *Image of Mars Rover Curiosity*

The instruments on board Curiosity enabled scientists to find out more about the surface of the planet.

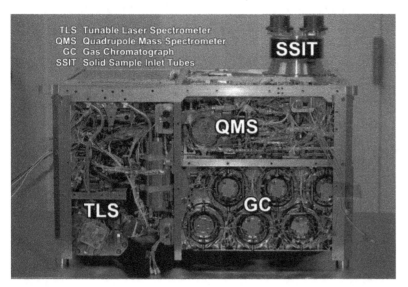

Figure 3.11.25: *SAM – the largest instrument on board Mars Curiosity Rover*

Sample Analysis at Mars (SAM), the largest instrument, is made up of three main components: a mass spectrometer, a gas chromatograph and a laser spectrometer.

SAM is being used to search for compounds of the element carbon, including methane, that are associated with life.

- *The mass spectrometer separates elements and compounds by mass for identification and measurement.*

- *The gas chromatograph heats soil and rock samples until they vaporise, and then separates the resulting gases into various components for analysis.*

- *The laser spectrometer measures the abundance of isotopes of carbon, hydrogen and oxygen.*

Quantitative analysis

Chemists are not only interested in what substances are present but how much of these substances are present. Finding out how much of a substance is present is termed **quantitative analysis**.

One method used to determine how much of a substance is present is titration.

Titration is used widely in industry. For example, in wineries it is used to check acidity as this affects the keeping quality of the wine and in the dairy industry it is part of a procedure that measures the protein content of foods.

Titration is also used to monitor the quality of water in lochs and rivers.

> 📖 **Word bank**
>
> • **Quantitative analysis**
> Detecting how much of a chemical is present in a substance.

Hint

Titration allows the concentration of the dissolved oxygen to be determined. The method is known as the Winkler method. An internet search will allow you find out more about this method.

Fish can't survive without good water quality. For trout and salmon the amount of oxygen in the water (known as dissolved oxygen) needs to be above 9mg l^{-1}. If it falls below this level the fish will become stressed and there may not be enough oxygen to keep them alive.

The water quality in many of Scotland's rivers has improved over recent years. Salmon are now being caught on the River Clyde after an absence from the river for many years. The Clyde suffered badly from industrial pollution.

Figure 3.11.26: *A salmon leaping on the River Tay*

How to do a titration

Titraton can be used to find the concentration of an acid or alkali using a neutralisation reaction.

The concentration of a sodium hydroxide solution can be found by titrating the solution with a dilute acid such as hydrochloric acid.

Word bank

• **Pipette**

Glass tube used to accurately measure and transfer an exact volume of solution.

Word bank

• **Indicator**

A chemical which changes colour at the end-point of a reaction.

Step 1   *A **pipette** is used to transfer a known quantity of the sodium hydroxide solution to a conical flask.*

The pipette should be rinsed with a small amount of the solution before use. The solution should be drawn up into the pipette until it is above the graduation mark on the pipette. It should then be allowed to slowly fall until the bottom of the meniscus just touches the mark. Make sure there are no air bubbles in the tip of the pipette as well.

Step 2   *Two or three drops of a suitable **indicator** are added to the solution and the flask placed on a white tile.*

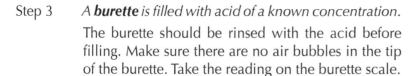

Step 3   *A **burette** is filled with acid of a known concentration.*

The burette should be rinsed with the acid before filling. Make sure there are no air bubbles in the tip of the burette. Take the reading on the burette scale.

Step 4   *Add the acid to the sodium hydroxide solution in the flask until there is a colour change.*

The acid should be added slowly and the conical flask swirled at the same time. As the end-point is approached the colour will begin to change. To obtain an accurate value the acid needs to be added dropwise at this point.

Concordant titres

When titration is used as an analytical tool, it is common practice to carry out a rough titre to estimate the volume of reagent that needs to be added to reach the end-point. The titration is then carried out accurately to find the precise volume that needs to be added. It is good practice to repeat the titration until concordant titres are obtained. **Concordant titres** are titres where the results are similar, usually within 0.2 cm³ of each other. The values are averaged (discarding the rough titration value) for use in a titration calculation (see page 85).

> 📖 **Word bank**
>
> • **Burette**
> Graduated glass tube with a tap at one end used to deliver variable volumes of a solution in titrations.

Figure 3.11.27: *Performing titration*

> 📖 **Word bank**
>
> • **Concordant titres**
> Titres volumes within 0.2 cm³ of each other.

Table 3.11.4: *Typical results for a titration*

Titration	Initial reading (cm³)	Final reading (cm³)	Volume added (cm³)
1 (rough)	0.3	15.5	15.2
2	15.2	29.9	14.7
3	29.9	44.4	14.5

The average titre is $\frac{(14.7 + 14.5)}{2} = 14.6$ cm³

Reporting experimental work

You should include the following, where appropriate, when writing a report about an experiment.

- Labelled sectional diagrams of common chemical apparatus.
- Tables of data with appropriate headings and units.
- Present data as a bar, line or scatter graph with suitable scales and labels.
- Draw a line of best fit to represent a trend observed in experimental data.
- Calculate an average (mean) from data.
- Suggest and justify an improvement to an experimental method.

Learning checklist

In this chapter you have learned:

- To recognise common pieces of apparatus and know how to use them.
- How to carry out common practical techniques:
    Filtration
    Using a balance
    Collecting gases
    Methods of heating
    Preparing soluble salts
    Preparing insoluble salts
    Testing electrical conductivity
    Setting up an electrochemical cell
    Electrolysis of solutions
    Determination of E_h.
- Qualitative analysis is the name given to the group of techniques that indicate which substances are present.
- Qualitative analysis techniques include flame testing, precipitation and gas tests.
- Quantitative analysis allows the amount of a substance present to be determined.
- Titration is a quantitative analysis method.
- How to carry out a titration.
- What to include in an experimental report.

Area 1: Chemical changes and structure

1.1 Rates of reaction

EXAM-STYLE QUESTIONS

1. Copper(II) carbonate reacts with dilute hydrochloric acid as shown.

$$CuCO_3(s) + 2HCl(aq) \rightarrow CuCl_2(aq) + H_2O(\ell) + CO_2(g)$$

A student used the apparatus shown below to follow the progress of the reaction.

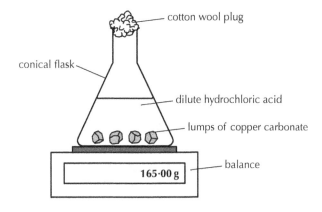

(a) Suggest why a cotton wool plug is placed in the mouth of the conical flask.

(b) The experiment was carried out using 0·50 g samples of both pure and impure copper(II) carbonate. The graph shows the results obtained.

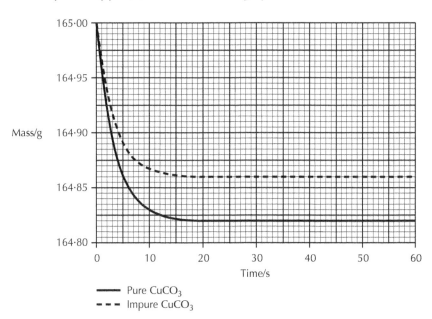

(i) For the sample of pure copper(II) carbonate, calculate the average reaction rate, in g s⁻¹, over the first 10 seconds.

(ii) Calculate the mass, in grams, of copper(II) carbonate present in the impure sample.
Show your working clearly.

2. Chloromethane, CH_3Cl, can be produced by reacting methanol solution with dilute hydrochloric acid using a solution of zinc chloride as a catalyst.

$$CH_3OH(aq) + HCl(aq) \rightarrow CH_3Cl(aq) + H_2O(\ell)$$

The graph shows how the concentration of the hydrochloric acid changed over a period of time when the reaction was carried out at 2°C.

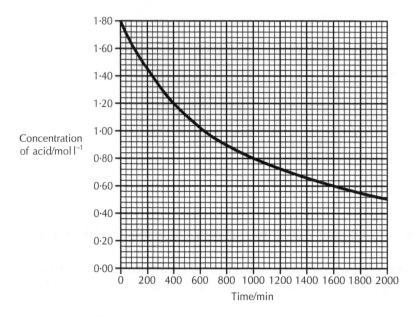

(a) Calculate the average rate of reaction for the first 400 minutes.
Your answer must include the appropriate unit.
Show your working clearly.

(b) On the graph above, sketch a curve to show how the concentration of hydrochloric acid would change over time if the reaction is repeated at 30°C.

3. Rapid inflation of airbags in cars is caused by the production of nitrogen gas. The graph gives information on the volume of gas produced over 30 microseconds.

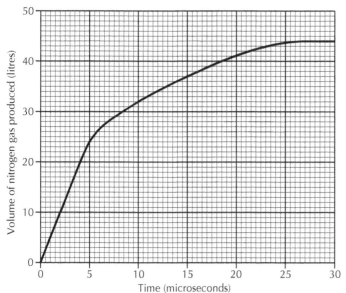

(a) (i) Calculate the average rate of reaction between 2 and 10 microseconds.
Your answer must include the appropriate unit.
Show your working clearly.

(ii) At what time has half of the final volume of nitrogen gas been produced?

4. Marble chips react with hydrochloric acid to produce carbon dioxide gas. The table shows the volume of carbon dioxide gas collected over time for the above reaction.

(a) Plot a line graph of the results of the reaction.

(b) Draw and label the apparatus and chemicals you would use to determine the volume of carbon dioxide gas produced when marble chips and hydrochloric acid react.

Time (seconds)	Volume of carbon dioxide (cm^3)
0	0
10	7
20	15
30	22
40	27
50	29
60	30
70	30

1.2 Atomic structure and bonding

EXAM-STYLE QUESTIONS

1. Isotopes of the same element have identical

A nuclei

B mass numbers

C numbers of neutrons

D numbers of protons.

2. Which of the following diagrams could be used to represent the structure of a covalent network compound?

A

B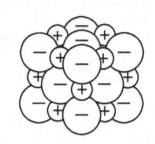

C

D

3. The table shows the colours of some ionic compounds in solution.

The colour of the chromate ion is

A colourless

B yellow

C green

D blue.

Compound	Colour
potassium chloride	colourless
potassium chromate	yellow
copper chromate	green
copper sulphate	blue

4. Information on some two-element molecules is shown in the table.

(a) Complete the table to show the **shape** of a molecule of ammonia.

(b) The hydrogen fluoride molecule can be represented as:

Name	Formula	Shape of molecule
hydrogen fluoride	HF	H-F
water	H_2O	H H
ammonia	NH_3	

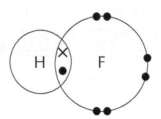

Showing **all** outer electrons, draw a similar diagram to represent a molecule of water, H_2O.

5. The element carbon can exist in the form of diamond.
    The structure of diamond is shown in the diagram.

    (a) Name the type of **bonding** and **structure** present in diamond.

    (b) Carbon forms many compounds with other elements such as hydrogen.
        (i) Draw a diagram to show how the outer electrons are arranged in a molecule of methane, CH_4.
        (ii) Draw a diagram to show the **shape** of a molecule of methane, CH_4.

6. Glass is made from the chemical silica, SiO_2, which is covalently bonded and has a melting point of 1700°C.

    (a) What does the melting point of silica suggest about its **structure**?

    (b) Antimony(III) oxide is added to reduce any bubbles that may appear during the manufacturing process.

        Write the chemical formula for antimony(III) oxide.

    (c) In the manufacture of glass, other chemicals can be added to alter the properties of the glass. The element boron can be added to glass to make ovenproof dishes.

        (i) Information about an atom of boron is given in the table below.

Particle	Number
proton	5
electron	5
neutron	6

        Use this information to complete the nuclide notation for this atom of boron.

        $$^{-}_{-}B$$

        (ii) Atoms of boron exist which have the same number of protons but a different number of neutrons from that shown in the table.
             What name can be used to describe the different atoms of boron?

7. The properties of a substance depend on its type of bonding and structure.
    There are four types of bonding and structure.

Covalent molecular	Covalent network
Ionic lattice	Metallic lattice

(a) Use these to complete the table, matching each type of bonding and structure with its properties.

Bonding and structure type	Properties
(a)	do not conduct electricity and have high melting points
(b)	have high melting points and conduct electricity when liquid but not when solid
(c)	conduct electricity when solid and have a wide range of melting points
(d)	do not conduct electricity and have low melting points

(b) A section of a covalent network compound is shown below.

● = silicon
○ = oxygen

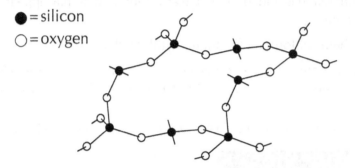

Write the formula for this covalent network compound.

1.3 Formulae and reaction quantities

EXAM-STYLE QUESTIONS

1. What is the name of the compound with the formula Ag_2O?

   A   Silver(I) oxide

   B   Silver(II) oxide

   C   Silver(III) oxide

   D   Silver(IV) oxide

2. The formula for magnesium sulphite is

   A   MgS

   B $MgSO_3$

   C $MgSO_4$

   D MgS_2O_3

3. $xAl(s) + yBr_2(\ell) \rightarrow zAlBr_3(s)$

This equation will be balanced when

A $x = 1, y = 2, z = 1$

B $x = 2, y = 3, z = 2$

C $x = 3, y = 2, z = 3$

D $x = 4, y = 3, z = 4$

4. (a) In some types of airbag, electrical energy causes sodium azide, NaN_3, to decompose producing sodium metal and nitrogen gas. The formula equation for the reaction is:

$$NaN_3(s) \rightarrow Na(s) + N_2(g)$$

Balance this equation.

(b) The equation for the decomposition of hydrogen peroxide is:

$$H_2O_2(aq) \rightarrow O_2(g) + H_2O(\ell)$$

Balance this equation.

5. (a) A strip of rhubarb was found to contain 1·8 g of oxalic acid.
How many moles of oxalic acid, $C_2H_2O_4$, are contained in 1·8 g?
(Formula mass of oxalic acid = 90)

(b) Permanganate ions (MnO_4^-) react with oxalic acid in rhubarb.
Calculate the number of moles of permanganate ions (MnO_4^-) in 100 cm³ of a 1·0 mol l⁻¹ solution.

6. Rust, iron(III) oxide, that forms on cars can be treated using rust remover which contains phosphoric acid.

The rust remover contains 250 cm³ of 2 mol l⁻¹ phosphoric acid.

Calculate the number of moles of phosphoric acid in the rust remover.

7. Silver jewellery slowly tarnishes in air. This is due to the formation of silver(I) sulfide.

The silver(I) sulfide can be converted back to silver by reacting it with aluminium.

The equation for the reaction which takes place is shown.

$$3Ag_2S(aq) + 2Al(s) \rightarrow 6Ag(s) + Al_2S_3(aq)$$

Calculate the mass of silver produced when 0·135 g of aluminium is used up.

1.4 Acids and bases

EXAM-STYLE QUESTIONS

1. Reactions can be represented using ionic equations. Which ionic equation shows a neutralisation reaction?

    A $2H_2O(\ell) + O_2(g) + 4e^- \rightarrow 4OH^-(aq)$

    B $H^+(aq) + OH^-(aq) \rightarrow H_2O(\ell)$

    C $SO_2(g) + H_2O(\ell) \rightarrow 2H^+(aq) + SO_3^{2-}(aq)$

    D $NH_4^+(s) + OH^-(s) \rightarrow NH_3(g) + H_2O(\ell)$

2. Which of the following oxides dissolves in water to produce a solution with a pH greater than 7?

    A Na_2O

    B Al_2O_3

    C SO_2

    D Ag_2O

3. Which line in the table describes what happens to a dilute solution of hydrochloric acid when water is added to it?

	pH	$H^+(aq)$ concentration
A	increases	increases
B	increases	decreases
C	decreases	increases
D	decreases	decreases

4. Which of the following statements describes the concentrations of $H^+(aq)$ and $OH^-(aq)$ ions in pure water?

    A The concentrations of $H^+(aq)$ and $OH^-(aq)$ ions are equal.

    B The concentrations of $H^+(aq)$ and $OH^-(aq)$ ions are zero.

    C The concentration of $H^+(aq)$ ions is greater than the concentration of $OH^-(aq)$ ions.

    D The concentration of $OH^-(aq)$ ions is greater than the concentration of $H^+(aq)$ ions.

5. Some indicators can have different colours when in solutions of different pH values.

   The tables give information about two indicators, bromothymol blue and methyl orange.

Bromothymol blue

Colour	pH
yellow	below 6·0
blue	above 7·6

Methyl orange

Colour	pH
red	below 3·1
yellow	above 4·4

The pH of three solutions was investigated using both indicators.

The results are shown below.

Substance	Colour with bromothymol blue	Colour with methyl orange
A	yellow	red
B	yellow	yellow
C	blue	yellow

(a) Which solution is alkaline?

(b) Suggest a pH value for solution B.

6. A solution of 0·1 mol l⁻¹ hydrochloric acid has a pH of 1.

   (a) (i) What colour would universal indicator turn when added to a solution of hydrochloric acid?
       (ii) Starting at pH 1, draw a line to show how the pH of this acid changes when diluted with water.

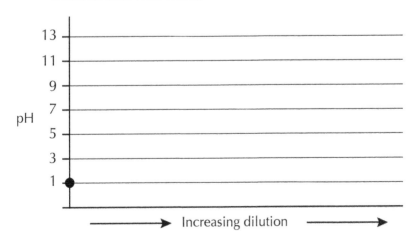

   (b) Calculate the number of moles of hydrochloric acid in 50 cm³ of 0·1 mol l⁻¹ hydrochloric acid solution.

7. The main use for sodium carbonate is glassmaking, for which a high purity is required. The purity of a sample of sodium carbonate can be checked by titration with acid.

$$Na_2CO_3(aq) + 2HCl(aq) \longrightarrow 2NaCl(aq) + CO_2(g) + H_2O(\ell)$$

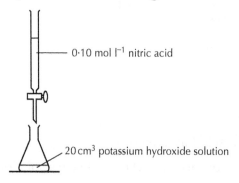

0·1 mol l^{-1} hydrochloric acid

25 cm^3 sodium carbonate solution

22·4 cm^3 acid was required in the titration of this sodium carbonate solution.

Calculate the concentration, in mol l^{-1}, of the sodium carbonate solution.

8. A student carried out a titration to find the concentration of a potassium hydroxide solution using 0·10 mol l^{-1} nitric acid.

0·10 mol l^{-1} nitric acid

20 cm^3 potassium hydroxide solution

    (a) What must be added to the flask to show the end-point of the titration?

    (b) The average volume of nitric acid needed to neutralise the potassium hydroxide solution is 24·6 cm^3.

    Calculate the concentration of the potassium hydroxide solution in mol l^{-1}.

Area 2: Nature's chemistry

2.5 Homologous series

EXAM-STYLE QUESTIONS

1. Which of the following compounds belongs to the same homologous series as the compound with the molecular formula C_3H_8?

A
```
    H   H
    |   |
H — C — C — H
    |   |
H — C — C — H
    |   |
    H   H
```

B
```
    H           H
    |           |
H — C — C = C — C — H
    |   |   |   |
    H   H   H   H
```

C
```
            H
            |
        H — C — H
    H   H   |   H
    |   |   |   |
H — C — C — C — C — H
    |   |   |   |
    H   H   H   H
```

D
```
            H
            |
        H — C — H
    H   H   |
    |   |   |
H — C — C — C = C — H
    |   |       |
    H   H       H
```

2.
```
        H
        |
    H — C — H
    H   |   H
    |   |   |
H — C — C — C — H
    |   |   |
    H   |   H
    H — C — H
        |
        H
```

Which of the following compounds is an isomer of the one shown above?

A
```
    H   H   H   H   H
    |   |   |   |   |
H — C — C — C = C — C — H
    |   |           |
    H   H           H
```

B
```
        H   H
         \ /
          C
    H     |     H
     \    |    /
      C       C
     /  \   /  \
    H    \ /    H
          |
      H — C — C — H
          |   |
          H   H
```

C
```
    H   H   H   H
    |   |   |   |
H — C — C — C — C — H
    |   |   |   |
    H   H   |   H
        H — C — H
            |
            H
```

D
```
            H
            |
        H — C — H
    H   H   |   H
    |   |   |   |
H — C — C — C — C — H
    |   |   |   |
    H   H   |   H
        H — C — H
            |
            H
```

3.

$$H-\underset{\underset{H}{|}}{\overset{\overset{H}{|}}{C}}-\underset{\underset{H}{|}}{\overset{\overset{H}{|}}{C}}-\underset{\underset{CH_3}{|}}{\overset{\overset{H}{|}}{C}}-\underset{\underset{H}{|}}{\overset{\overset{CH_3}{|}}{C}}-\underset{\underset{H}{|}}{\overset{\overset{H}{|}}{C}}-H$$

The name of the above compound is

A   2,3-dimethylpentane

B   3,4-dimethylpentane

C   2,3-dimethylpropane

D   3,4-dimethylpropane

4.   Pentane can be converted to 2-methylbutane for blending into petrol by a process called isomerisation. Part of the sequence of reactions in the conversion is shown below.

pentane → pent-2-ene → 2-methylbut-2-ene → 2-methylbutane

(a) Suggest a name for the type of reaction which converts pentane to pent-2-ene.

(b) Draw a structural formula for 2-methylbut-2-ene.

(c) Why can the process be called isomerisation?

(d) Why are branched-chain hydrocarbons blended into petrol?

5.   Alkenes can be made from bromoalkanes. Bromoalkanes are alkane molecules in which a hydrogen atom has been replaced by a bromine atom.

$$H-\underset{\underset{H}{|}}{\overset{\overset{H}{|}}{C}}-\underset{\underset{H}{|}}{\overset{\overset{H}{|}}{C}}-\underset{\underset{H}{|}}{\overset{\overset{H}{|}}{C}}-\underset{\underset{Br}{|}}{\overset{\overset{H}{|}}{C}}-H \quad \rightarrow \quad H-\underset{\underset{H}{|}}{\overset{\overset{H}{|}}{C}}-\underset{\underset{H}{|}}{\overset{\overset{H}{|}}{C}}-\overset{\overset{H}{|}}{C}=\overset{\overset{H}{|}}{C}-H \quad + \quad HBr$$

1-bromobutane   but-1-ene

Draw the full structural formula for the **two** alkenes which can be formed from 2-bromobutane.

$$H-\underset{\underset{H}{|}}{\overset{\overset{H}{|}}{C}}-\underset{\underset{H}{|}}{\overset{\overset{H}{|}}{C}}-\underset{\underset{Br}{|}}{\overset{\overset{H}{|}}{C}}-\underset{\underset{H}{|}}{\overset{\overset{H}{|}}{C}}-H$$

2-bromobutane

6.   (a) Name the alkane shown below.

$$H-\underset{\underset{H}{|}}{\overset{\overset{H}{|}}{C}}-\underset{\underset{CH_3}{|}}{\overset{\overset{H}{|}}{C}}-\underset{\underset{H}{|}}{\overset{\overset{CH_3}{|}}{C}}-\underset{\underset{H}{|}}{\overset{\overset{H}{|}}{C}}-H$$

(b) Alkanes can be reacted with alkenes to produce longer chain alkanes.

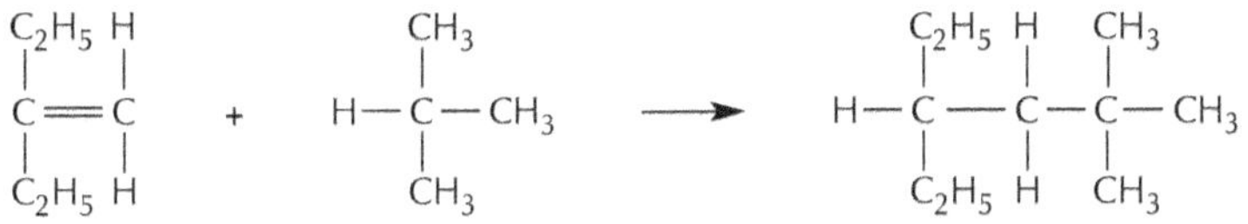

$$
\underset{\begin{array}{c}C_2H_5\ \ H\\|\quad\ |\\C=C\\|\quad\ |\\C_2H_5\ \ H\end{array}}{}
\quad + \quad
\underset{\begin{array}{c}CH_3\\|\\H-C-CH_3\\|\\CH_3\end{array}}{}
\quad \longrightarrow \quad
\underset{\begin{array}{c}C_2H_5\ \ H\ \ \ CH_3\\|\quad\ |\quad\ |\\H-C-C-C-CH_3\\|\quad\ |\quad\ |\\C_2H_5\ \ H\ \ \ CH_3\end{array}}{}
$$

Draw the structural formula of the alkane formed in the following reaction.

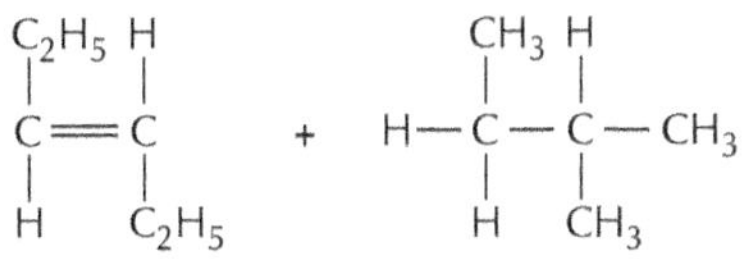

$$
\underset{\begin{array}{c}C_2H_5\ \ H\\|\quad\ |\\C=C\\|\quad\ |\\H\quad\ C_2H_5\end{array}}{}
\quad + \quad
\underset{\begin{array}{c}CH_3\ \ H\\|\quad\ |\\H-C-C-CH_3\\|\quad\ |\\H\quad\ CH_3\end{array}}{}
$$

7. Gases can be liquefied by increasing the pressure, but above a certain temperature it is not possible to do this. This temperature is known as the critical temperature. The critical temperatures of some alkanes are shown below.

Alkane	Critical temperature (°C)
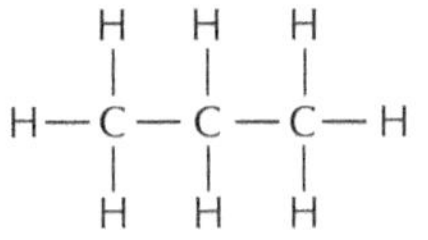	97
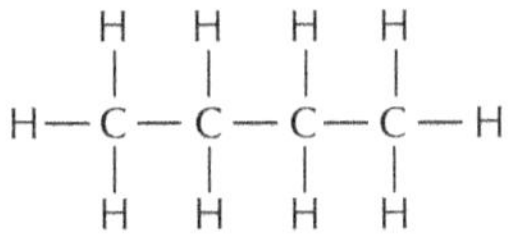	152
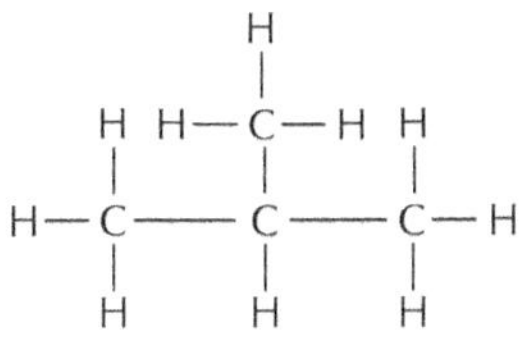	135
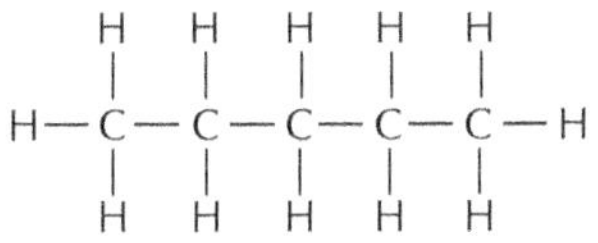	197
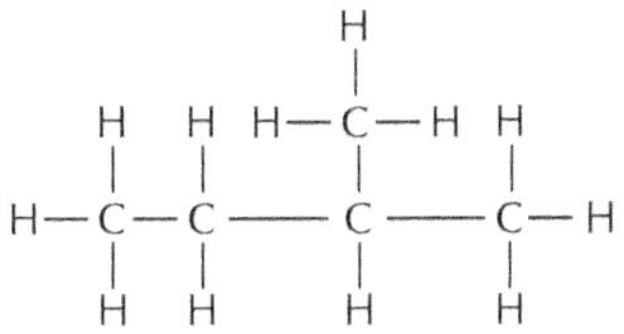	187
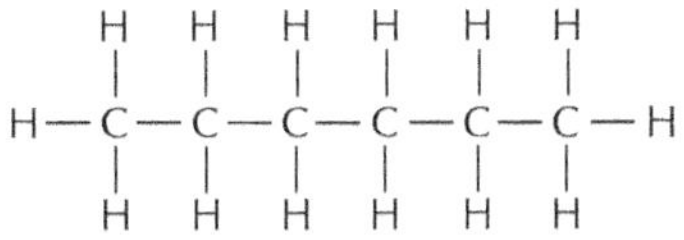	234

(a) Describe the trend in critical temperatures for the straight-chain alkanes.

(b) Predict the critical temperature of this alkane.

2.6 Everyday consumer products

EXAM-STYLE QUESTIONS

1. Alcohols are useful solvents.

   (a) Name the functional group in all alcohols.

   (b) (i) Draw a structural formula for propan-1-ol.

   (ii) Explain fully why hexan-1-ol has the highest boiling point of these alcohols.

   (iii) Predict the boiling point of octan-1-ol.

   (c) Draw the full structural formula for an isomer of propan-1-ol.

Alcohol	Boiling point (°C)
propan-1-ol	97
butan-1-ol	118
pentan-1-ol	138
hexan-1-ol	158

2. The thiols are a family of compounds containing carbon, hydrogen and sulfur.

   (a) Thiols have the same general formula and similar chemical properties.

   (i) State the term used to describe a family of compounds such as the thiols.

   (ii) Suggest a general formula for this family.

   (b) Ethanethiol can react with oxygen as shown.

   ethanethiol + oxygen → carbon dioxide + water + Y
   Identify Y.

   (c) Draw an isomer of propanethiol.

Name	Full structural formula
methanethiol	
ethanethiol	
propanethiol	

3. The table gives information about the amount of energy released when 1 mole of some alcohols are burned.

(a) Draw a structural formula for pentan-2-ol.

(b) (i) Write a statement linking the amount of energy released to the position of the functional group in an alcohol molecule.

(ii) Predict the amount of energy released, in KJ, when 1 mole of hexan-2-ol is burned.

Name of alcohol	Energy released when 1 mole of alcohol is burned (KJ)
propan-1-ol	2021
propan-2-ol	2005
butan-1-ol	2676
butan-2-ol	2661
pentan-1-ol	3329
pentan-2-ol	3315
hexan-1-ol	3984

4. Alkanoic acids are a family of compounds which contain the

group.

The **full** structural formulae for the first three members are shown.

(a) Draw the **full** structural formula for the alkanoic acid containing 4 carbon atoms.

(b) The table gives information on some alkanoic acids.

Acid	Boiling point/°C
methanoic acid	101
ethanoic acid	118
propanoic acid	141
butanoic acid	164

(i) Using this information, make a general statement linking the boiling point to the number of carbon atoms.

(ii) Predict the boiling point of pentanoic acid.

Area 3: Chemistry in society

3.7 Metals

EXAM-STYLE QUESTIONS

1. Although they are more expensive, fuel cells have been developed as an alternative to petrol for motor vehicles.

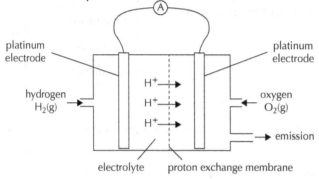

(a) (i) The ion-electron equations for the process occurring at each electrode are:

$$H_2(g) \rightarrow 2H^+(aq) + 2e^-$$

$$O_2(g) + 4H^+(aq) + 4e^- \rightarrow 2H_2O(\ell)$$

Combine these two equations to give the overall redox equation.

(ii) On the diagram, show by means of an arrow the path of electron flow.

(b) Suggest one advantage that fuel cells have over petrol for providing energy.

2. Fuel cells can be used to power cars.

The ion-electron equations for the oxidation and reduction reactions that take place in a methanol fuel cell are:

$$CH_3OH(\ell) + H_2O(\ell) \rightarrow CO_2(g) + 6H^+(aq) + 6e^-$$

$$3O_2(g) + 12H^+(aq) + 12e^- \rightarrow 6H_2O(\ell)$$

(a) Combine the two ion-electron equations to give the equation for the overall redox reaction.

(b) The equation for the overall redox reaction in a hydrogen fuel cell is

$$2H_2(g) + O_2(g) \rightarrow 2H_2O(\ell)$$

Give a disadvantage of the methanol fuel cell reaction compared to the hydrogen fuel cell reaction.

3. Titanium is an important metal.

Titanium can be extracted from titanium dioxide. The titanium dioxide is reacted with carbon and chlorine to produce impure titanium chloride and carbon dioxide. The impure titanium chloride is purified by distillation. Magnesium metal is added to the pure titanium chloride producing titanium and magnesium chloride.

Complete the flow chart to show the extraction process.

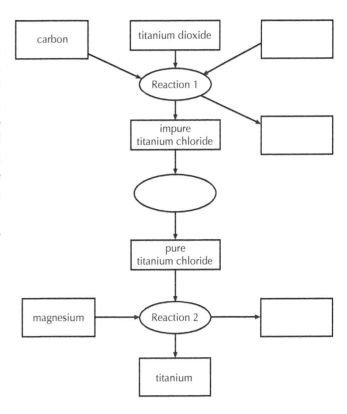

4. A mixture of titanium and nickel is used to make the alloy, Nitinol. This alloy is used to make dental braces. The composition of Nitinol is shown in the table.

Metal	titanium	nickel
Percentage by mass	45	55

A set of braces has a mass of 8 g.

(a) Calculate the mass of titanium in the braces.

(b) Calculate the number of moles of titanium in the braces.
(Take the relative atomic mass of titanium to be 48.)

3.8 Plastics

EXAM-STYLE QUESTIONS

1. The structure below shows a section of an addition polymer.

$$\begin{array}{ccccccc} & H & & CH_3 & & H & & CH_3 & & H & & CH_3 \\ & | & & | & & | & & | & & | & & | \\ \text{---} & C & \text{---} & C & \text{---} & C & \text{---} & C & \text{---} & C & \text{---} & C & \text{---} \\ & | & & | & & | & & | & & | & & | \\ & H & & COOCH_3 & & H & & COOCH_3 & & H & & COOCH_3 \end{array}$$

Which molecule is used to make this polymer?

A
$$CH_3 \quad H$$
$$| \qquad |$$
$$C = C$$
$$| \qquad |$$
$$H \qquad COOCH_3$$

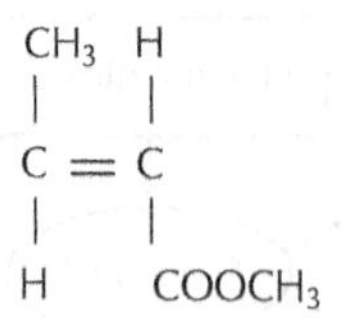

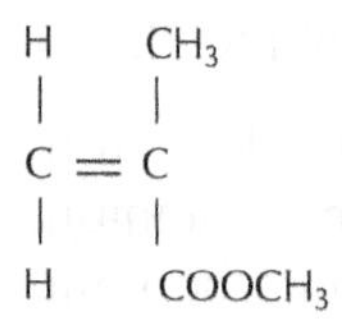

B
$$H \qquad CH_3$$
$$| \qquad |$$
$$C = C$$
$$| \qquad |$$
$$H \qquad COOCH_3$$

C
$$CH_3 \quad COOCH_3$$
$$| \qquad |$$
$$C = C$$
$$| \qquad |$$
$$H \qquad H$$

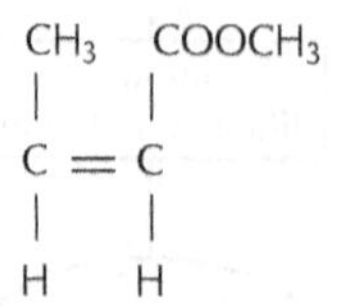

D
$$H \qquad CH_3$$
$$| \qquad |$$
$$H - C - C - H$$
$$| \qquad |$$
$$H \qquad COOCH_3$$

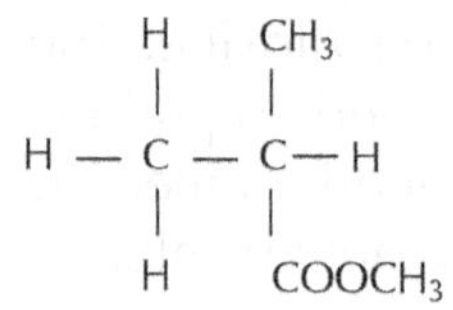

2. Part of the structure of an addition polymer is shown below. It is made using two different monomers.

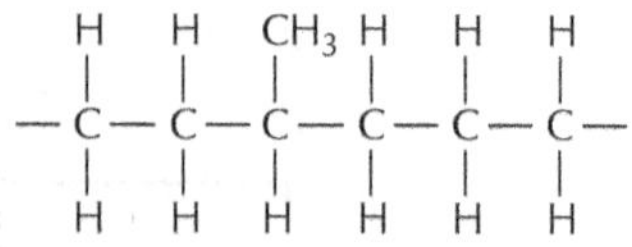

$$H \quad H \quad CH_3 \; H \quad H \quad H$$
$$| \quad | \quad | \quad | \quad | \quad |$$
$$- C - C - C - C - C - C -$$
$$| \quad | \quad | \quad | \quad | \quad |$$
$$H \quad H \quad H \quad H \quad H \quad H$$

Which pair of alkenes could be used as monomers for this polymer?

A Ethene and propene

B Ethene and butene

C Propene and butene

D Ethene and pentene

3. Poly(ethenol) is one of the substances used to cover dishwasher tablets. A section of the poly(ethenol) polymer is shown.

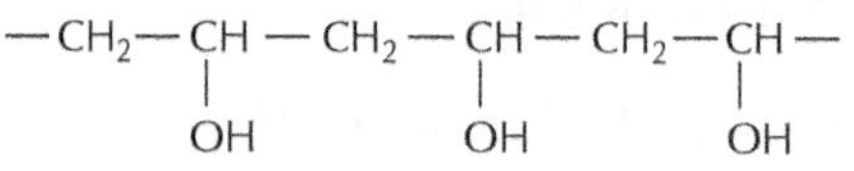

$$- CH_2 - CH - CH_2 - CH - CH_2 - CH -$$
$$| \qquad\qquad | \qquad\qquad |$$
$$OH \qquad\quad OH \qquad\quad OH$$

(a) Name the functional group present in this polymer.

(b) Draw the structure of the repeating unit for this polymer.

(c) A dishwasher tablet, complete with its poly(ethenol) cover, can be added to a dishwasher.

What property of the poly(ethenol) makes it suitable as a cover for a dishwasher tablet?

4. Polystyrene is made from the monomer, styrene. The systematic name for styrene is phenylethene.

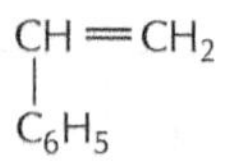

$$CH = CH_2$$
$$|$$
$$C_6H_5$$

styrene (phenylethene)

(a) The monomer used to form polystyrene is shown.

Which part of the structure of styrene allows the polymer to form?

(b) Complete the diagram to show how three styrene molecules join to form part of the polymer chain.

$$\sim \overset{|}{\underset{|}{C}} - \overset{|}{\underset{|}{C}} - \overset{|}{\underset{|}{C}} - \overset{|}{\underset{|}{C}} - \overset{|}{\underset{|}{C}} - \overset{|}{\underset{|}{C}} \sim$$

(c) Give another name for polystyrene.

5. The monomer in superglue has the following structure.

$$\begin{array}{cc} H & COOCH_3 \\ | & | \\ C & = C \\ | & | \\ H & CN \end{array}$$

Draw a section of the polymer, showing **three** monomer units joined together.

6. Poly(methyl methacrylate) is a synthetic polymer used to manufacture perspex.

(a) What is meant by the term **synthetic**?

(b) The structure of the methyl methacrylate monomer is shown.

$$\begin{array}{cc} H & CH_3 \\ | & | \\ C & = C \\ | & | \\ H & COOCH_3 \end{array}$$
methyl methacrylate

(i) Draw a section of the poly(methyl methacrylate) polymer, showing three monomer units joined together.

(ii) Name the type of polymerisation taking place.

3.9 Fertilisers

EXAM-STYLE QUESTIONS

1. Potassium hydroxide reacts with sulfuric acid to form potassium sulfate, which can be used as a fertiliser.

$$KOH(aq) + H_2SO_4(aq) \rightarrow K_2SO_4(aq) + H_2O(\ell)$$

(a) Balance the above equation.

(b) Name the type of chemical reaction taking place.

(c) Calculate the percentage, by mass, of potassium in potassium sulphate, K_2SO_4.
Show your working clearly.

(d) Ammonium phosphate is also used as a fertiliser.
Write the **ionic** formula for ammonium phosphate.

2. (a) The flow diagram shows how ammonia is converted to nitric acid.

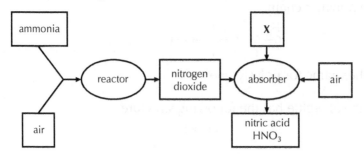

(i) Name the industrial process used to manufacture nitric acid.

(ii) The reactor contains a platinum catalyst.
Why is it **not** necessary to continue heating the catalyst once the reaction has started?

(iii) Name substance **X**.

(b) Ammonia and nitric acid react together to form ammonium nitrate, NH_4NO_3.

Calculate the percentage by mass of nitrogen in ammonium nitrate.

Show your working clearly.

3. Nitrogen is essential for healthy plant growth.

Nitrogen from the atmosphere can be fixed in a number of ways.

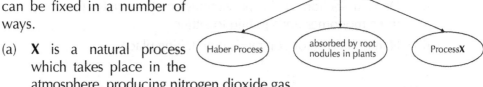

(a) **X** is a natural process which takes place in the atmosphere, producing nitrogen dioxide gas.

Suggest what provides the energy for this process.

(b) The Haber process is the industrial method of converting nitrogen into a nitrogen compound.

Name the nitrogen compound produced.

(c) The nitrogen compound produced in the Haber process dissolves in water.

The graph shows the solubility of the nitrogen compound at different temperatures.

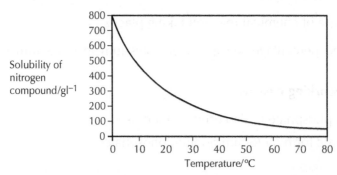

Write a general statement describing the effect of temperature on the solubility of the nitrogen compound.

4. Ammonia is made when nitrogen and hydrogen react together.

    The table below shows the percentage yields obtained when nitrogen and hydrogen react at different pressures.

Pressure/atmospheres	Percentage yield of ammonia
25	28
50	40
100	53
200	67
400	80

    (a) Draw a line graph of percentage yield against pressure.

    Use appropriate scales to fill most of the graph paper.

    (b) Use your graph to estimate the percentage yield of ammonia at 150 atmospheres.

    (c) Ammonia can be produced in the lab by heating an ammonium compound with soda lime.

    In order to produce ammonia, what **type** of compound must soda lime be?

5. Ammonia gas is produced when barium hydroxide reacts with ammonium chloride.

    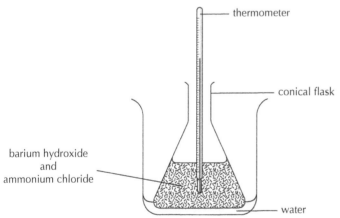

    (a) The equation for the reaction which takes place is:

    $Ba(OH)_2 + NH_4Cl \rightarrow NH_3 + BaCl_2 + H_2O$

    Balance this equation.

    (b) Describe a test which would detect ammonia at the mouth of the flask.

    (c) During the reaction the reading on the thermometer dropped from 25°C to −5°C.

    Suggest what would happen to the water in the beaker.

3.10 Nuclear chemistry

EXAM-STYLE QUESTIONS

1. The diagram shows the paths of alpha, beta and gamma radiations as they pass through an electric field.

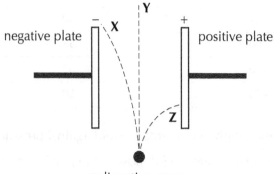

Which line in the table correctly identifies the types of radiation which follow paths **X**, **Y** and **Z**?

	Path X	Path Y	Path Z
A	gamma	beta	alpha
B	beta	gamma	alpha
C	beta	alpha	gamma
D	alpha	gamma	beta

2. An atom of ^{227}Th decays by a series of alpha emissions to form an atom of ^{211}Pb.

   How many alpha particles are released in the process?

   A 2

   B 3

   C 4

   D 5

3. The half-life of the isotope ^{210}Pb is 21 years. What fraction of the original ^{210}Pb atoms will be present after 63 years?

   A 0·5

   B 0·25

   C 0·125

   D 0·0625

4. Some smoke detectors make use of radiation which is very easily stopped by tiny smoke particles moving between the radioactive source and the detector.

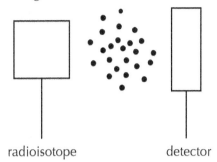

radioisotope   detector

The most suitable type of radioisotope for a smoke detector would be

A   an alpha-emitter with a long half-life

B   a gamma-emitter with a short half-life

C   an alpha-emitter with a short half-life

D   a gamma-emitter with a long half-life.

5. Tritium is an isotope of hydrogen. It is formed in the upper atmosphere when neutrons from cosmic rays are captured by nitrogen atoms.

$$^{14}_{7}N + ^{1}_{0}n \rightarrow ^{12}_{6}C + ^{3}_{1}H$$

Tritium atoms then decay by beta-emission.

$$^{3}_{1}H \rightarrow \qquad +$$

(a)   Complete the nuclear equation above for the beta-decay of tritium atoms.

(b)   In the upper atmosphere, tritium atoms are present in some water molecules. Over the years, the concentration of tritium atoms in rain has remained fairly constant.

(i)   Why does the concentration of tritium in rain remain fairly constant?

(ii) The concentration of tritium atoms in fallen rainwater is found to decrease over time. The age of any product made with water can be estimated by measuring the concentration of tritium atoms.

In a bottle of wine, the concentration of tritium atoms was found to be 1/8 of the concentration found in rain.

Given that the half-life of tritium is 12·3 years, how old is the wine?

6. All the isotopes of technetium are radioactive.

(a)   Technetium-99m is produced as shown.

$$^{99}_{42}Mo \rightarrow ^{99}_{43}Tc + X$$

Identify X.

(b) The graph shows the decay curve for a 1·0 g sample of technetium-99m.

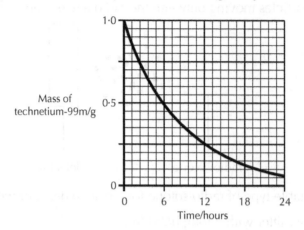

(i) Draw a curve on the graph to show the variation of mass with time for a 0·5 g sample of technetium-99m.

(ii) Technetium-99m is widely used in medicine to detect damage to heart tissue. It is a gamma-emitting radioisotope and is injected into the body.

Suggest **one** reason why technetium-99m can be safely used in this way.

7. Phosphorus-32 and strontium-89 are two radioisotopes used to study how far mosquitoes travel.

(a) Strontium-89 decays by emission of a beta particle.

Complete the nuclear equation for the decay of strontium-89.

$$^{89}Sr \rightarrow$$

(b) In an experiment, 10 g of strontium-89 chloride was added to a sugar solution used to feed mosquitoes.

(i) The strontium-89 chloride solution was fed to the mosquitoes in a laboratory at 20°C. When the mosquitoes were released, the outdoor temperature was found to be 35°C.

What effect would the increase in temperature have on the half-life of the strontium-89?

(ii) Calculate the mass, in grams, of strontium-89 present in the 10 g sample of strontium-89 chloride, $SrCl_2$.

(c) A mosquito fed on a solution containing phosphorus-32 is released.

Phosphorus-32 has a half-life of 14 days.

When the mosquito is recaptured 28 days later, what fraction of the phosphorus-32 will remain?

8.

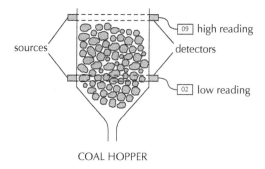

COAL HOPPER

An important industrial use of radioisotopes is in level gauges.

In industrial plants the height of coal in hoppers can be monitored by placing high energy gamma sources at various heights along one side of the hoppers. The radiation from the sources is focused on detectors placed opposite the sources.

(a) How are gamma radiations produced?

(b) Explain how this system will indicate the height of coal in the hopper.

(c) Suggest why a gamma source is used in these types of gauges.

9. Polonium-210 is a radioisotope that decays by alpha-emission.

The half-life of polonium-210 is 140 days.

(a) Draw a graph to show how the mass of 200 g of the radioisotope would change

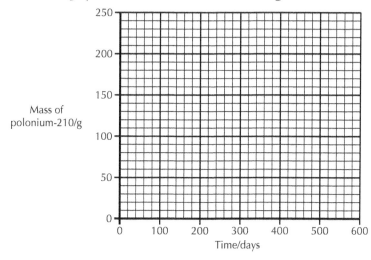

(b) Write a balanced nuclear equation for the alpha-decay of polonium-210.

3.11 Chemical analysis

EXAM-STYLE QUESTIONS

1. Carbon dioxide

    A burns with a pop

    B relights a glowing splint

    C turns damp pH paper blue

    D turns limewater cloudy.

2. Which of the following is the best way to collect a pure sample of a soluble gas which is less dense than air?

    A Downward displacement of water.

    B Gas syringe.

    C Downward displacement of air.

    D Upward displacement of air.

3. The work card details how the volume of acid needed to neutralise an alkali can be found. Steps 2 and 6 are missing.

    > **Neutralisation work card**
    >
    > Method
    >
    > 1 Transfer exactly 20 cm³ of the alkali to a conical flask.
    >
    > 2
    >
    > 3 Add acid to the burette and note the reading.
    >
    > 4 Add acid to the alkali while swirling the conical flask.
    >
    > 5 Stop adding the acid when the end-point is reached.
    >
    > 6

    (a) Complete the work card by giving instructions for steps 2 and 6.

    (b) Name the piece of glassware used to transfer the alkali to the conical flask in step 1.

    (c) What name is given to the analytical technique detailed in the work card?

    (d) The results of the experiments carried out are shown in the table.

Experiment	Volume of acid added (cm³)
1	19.8
2	19.5
3	19.3

    Calculate the volume of acid which would be used to work out the concentration of the acid.

Open-ended and extended response questions

Open-ended questions

In **open-ended** questions there is **no one specific answer**. There are many different ways of answering the question and they could all be correct. Open-ended questions give you the opportunity to let the examiner know how well you **understand** the area the question relates to without expecting a set answer. These questions give you credit for being **creative** and **analytical** in your answers.

Note that open-ended questions are worth a total of three marks, which indicates that quite a lot of information is needed in the answer. It doesn't mean that you need three correct points to gain all three marks but, the more relevant information you give and how well your answer is put together, the more marks you will get. The examiner wants to see that you have a **good understanding**.

You will recognise an open-ended question by the wording 'Using your knowledge of chemistry, comment on … '.

Open-ended questions should not take more than **five minutes** to answer and there are unlikely to be more than two open-ended questions in an exam paper.

Example 1

Iodine-131 is a radioisotope used in hospital. Some health workers who handle linen which has been in contact with the iodine-131 have expressed some concern about it.

From your knowledge of chemistry, comment on why health workers may be concerned about the use of iodine-131.

(3 marks)

🔵 GO! Paired activity

Work with a partner and discuss what you think would be an answer which would be considered to show good understanding. You may have to look at the section in the book that describes radioactivity and radioisotopes.

A sample answer can be found on page 280 – remember your answer may still show good understanding even if it doesn't include everything covered in the sample answer. Ask your teacher for their opinion.

Example 2

A student was measuring the pH of various solutions and concluded that only acids contained hydrogen ions and only alkalis contained hydroxide ions.

Think about the student's conclusion, discuss it with a partner and using your knowledge of chemistry, comment on the statement made.

A sample answer can be found on page 280.

> **GO!** **Paired activity**
>
> Work with a partner and discuss what you think would be an answer which would be considered to show good understanding. You may have to look at the section in the book on acids and alkalis.
>
> A sample answer can be found on page 280. Remember your answer may still show good understanding even if it doesn't include everything covered in the sample answer. Ask your teacher for their opinion.

Extended response questions

An **extended response** question is different from an open-ended question in that it is looking for specific points to explain what has been described in the question. The answer is worth two marks so two relevant points are normally expected in your answer.

Example

Smoke alarms are usually found attached to the ceiling in houses.

They contain the radioisotope americium-241 which emits α particles. Despite this, they are considered to be safe for use in the home.

Explain why the presence of americium-241 is not considered a health risk.

(2 marks)

> **GO!** **Paired activity**
>
> Work with a partner and discuss what you would consider to be a good explanation. You may have to look at the section in the book which describes the properties of radioactive emissions.
>
> A sample answer can be found on page 280.

Answers to activities

Area 1: Chemical changes and structure

1.1 Rates of reaction

GO! Activity 1.1.1

1. Presence of water could slow the reaction down (or the reaction may not happen) and sodium reacts violently with water.
2. The smaller the particle size, the bigger the surface area, the faster the reaction.
3. Sodium is very reactive and can burst into flames.

GO! Activity 1.1.2

1. (a) Weigh out powdered zinc.
   Add to flask.
   Add a measured volume of acid to flask.
   Start stop clock.
   Collect gas by displacement of water or in gas syringe.
   Note volume of gas collected every minute.
   Either of the arrangements shown can be used.

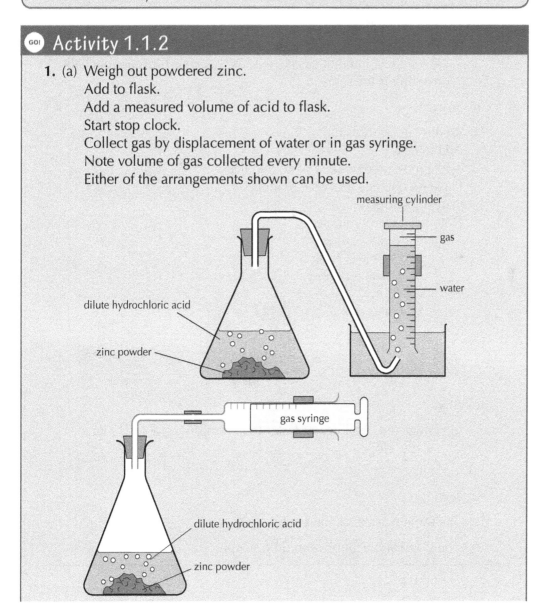

(b) (i) A

    (ii) Steeper at A so reaction is faster.

    (iii) Stopped

(c) 30 cm³

(d) (i)

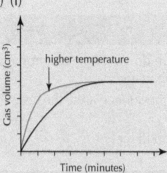

    (ii) The slope is steeper near the start because the reaction is faster. The final volume of gas is the same because the mass of chalk reacting is the same.

(e) 34 cm³

2. (a) Weigh out marble chips.
Add to flask.
Add a measured volume of acid to flask.
Place on balance and note reading.
Start stop clock.
Note mass every minute.

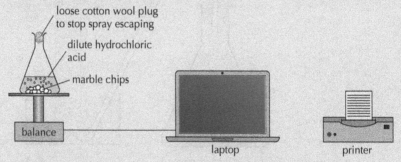

(b) (i) X

    (ii) Powdered chalk reacts faster than lumps so graph will have a steeper slope.

(c) 1·5 min

(d) 0·8 g

(e) The same mass of chalk reacts.

(f) The slope of the graph would be steeper.

GO! Activity 1.1.3

1. (a) Average rate of reaction $= \dfrac{\text{volume at 40s} - \text{volume at 20s}}{\text{time interval}}$

$$= (76-49)/(40-20)$$
$$= 1 \cdot 75 \text{ cm}^3 \text{ s}^{-1}$$

(b) The average rate is decreasing as the reactants are being used up.

(c) 62 ($\pm$2)s – the graph levels out showing no more gas is being produced.

2. Average rate of reaction $= \dfrac{\text{mass lost at 10s} - \text{mass lost at 0s}}{\text{time interval}}$

$$= (0 \cdot 05 - 0)/(10 - 0)$$
$$= 0 \cdot 05/10$$
$$= 0 \cdot 005 \text{ g s}^{-1}$$

3. Average rate of reaction $= \dfrac{\text{concentration at 20s} - \text{concentration at 0s}}{\text{time interval}}$

$$= (0 \cdot 64 - 0)/(20 - 0)$$
$$= 0 \cdot 64/20$$
$$= 0 \cdot 032 \text{ mol l}^{-1} \text{ s}^{-1}$$

4. (a)

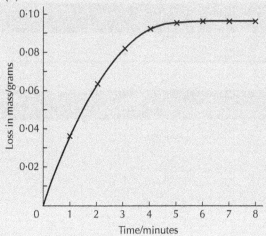

(b) The reaction has stopped because at least one of the reactants has been completely used up.

1.2 Atomic structure and bonding

GO! Activity 1.2.1

(a)	(a) protons	(b) (n)	(c) electrons	(d) neutrons
	(e) nucleus	(f) electrons	(g) positive	(h) 1
	(i) no	(j) negative	(k) zero	

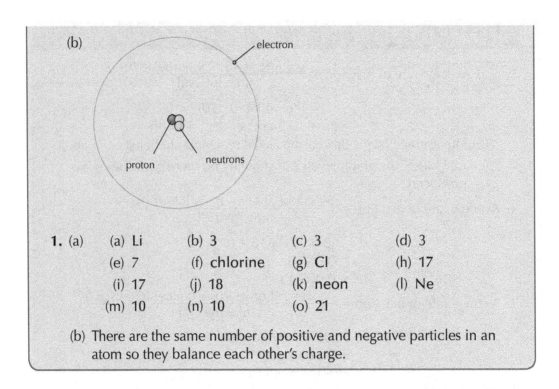

(b)

1. (a)

(a) Li	(b) 3	(c) 3	(d) 3
(e) 7	(f) chlorine	(g) Cl	(h) 17
(i) 17	(j) 18	(k) neon	(l) Ne
(m) 10	(n) 10	(o) 21	

(b) There are the same number of positive and negative particles in an atom so they balance each other's charge.

GO! Activity 1.2.2

1. (a) $^{27}_{13}\text{Al}$ (b) $^{27}_{13}\text{Al}^{3+}$

GO! Activity 1.2.3

1. RAM = $[(28 \times 92) + (29 \times 5) + (30 \times 3)]/100$

RAM = 28·1

2. (a)

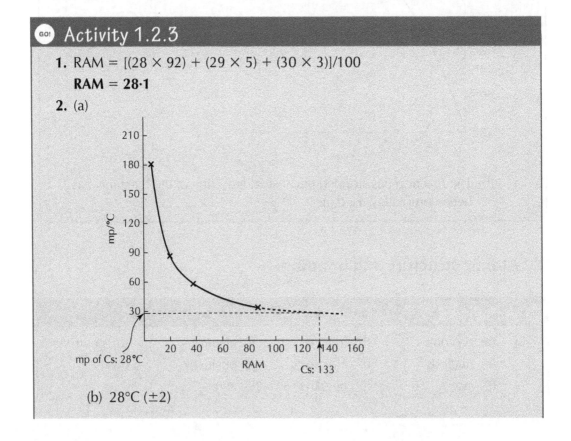

(b) 28°C (±2)

3. (a) **mass spectrometer**: instrument for measuring the mass of atoms

isotopes: atoms with the same number of protons but different numbers of neutrons

mass number: number of protons + neutrons in an atom

relative atomic mass: average mass of the atoms of an element taking into consideration the isotopes

(b) (i) Three – there are three peaks

(ii) 24 (80%)

(iii) Relative atomic mass is the average mass of the three isotopes taking into account the percentage of each.

GO! Activity 1.2.4

1. (a)

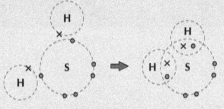

(b) H_2S : hydrogen sulfide

(c) Angular

(d) Oxygen and sulfur are both in Group 6 so have 6 electrons in their outer energy level.

2. (a)

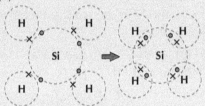

(b) SiH_4 : silicon hydride (silane)

(c) Tetrahedral

(d) Silicon and carbon are both in Group 4 so have 4 electrons in their outer energy level.

Activity 1.2.5

1. (a) individual
   (b) networks
   (c) liquid
   (d) covalent
   (e) strong
   (f) weak
   (g) energy
   (h) separate
   (i) low
   (j) high
   (k) covalently
   (l) three-dimensional
   (m) bonds

2. Carbon dioxide is molecular so has weak forces of attraction between molecules that do not require a lot of energy to separate them. Silicon dioxide is a covalent network with strong covalent bonds holding the atoms together, which requires a lot of energy to break the bonds.

Activity 1.2.6

1. (a) A = covalent molecular
      B = covalent network
      C = ionic

   (b) Covalent substances do not conduct electricity in any state so A and B must be covalent. Network substances have high melting and boiling points because of the strong covalent bonds between atoms, so B must be covalent network. A must be covalent molecular because of its low melting and boiling points. C must be ionic because only ionic substances conduct electricity in solution or when liquid.

   (c) C is ionic and the ions are not free to move in the solid state but can move in solution and as a liquid.

2. Ionic compounds can be coloured and are always solid. Covalent substances can be white solids and colourless liquids or gases.

3. (a) Any of the compounds could be selected because they are all ionic solids.

   (b) Lithium chloride is a good choice because it has the lowest melting point so is the easiest to melt.

   (c) Calcium carbonate is a bad choice because it is insoluble.

1.3 Formulae and reaction quantities

GO! Activity 1.3.1

(a) Mg (b) Cl_2 (c) $BH_3(H_3B)$ (d) CO (e) PCl_3

GO! Activity 1.3.2

(a) hydrogen + bromine → hydrogen bromide

$\quad\quad H_2 \quad + \quad Br_2 \quad → \quad\quad HBr$

(b) phosphorus + chlorine → phosphorus pentachloride

$\quad\quad P \quad + \quad Cl_2 \quad → \quad PCl_5$

GO! Activity 1.3.3

1. (a) Na_2SO_3 (b) K_2SO_4 (c) $Ca(HCO_3)_2$ (d) $Co(NO_3)_3$ (e) $(NH_4)_3PO_4$
(f) $Mg(OH)_2$

2. (a) zinc + copper(II) sulfate → copper + zinc sulfate

$\quad Zn(s) + \quad CuSO_4(aq) \quad → Cu(s) + ZnSO_4(aq)$

(b) barium + lithium → barium + lithium
   hydroxide   sulfate   sulfate   hydroxide

$\quad Ba(OH)_2(aq) + Li_2SO_4(aq) → BaSO_4(s) + LiOH(aq)$

GO! Activity 1.3.4

1. $2K + S → K_2S$
2. $2Na + Cl_2 → 2NaCl$
3. $C + CO_2 → 2CO$
4. $2AgNO_3 + MgCl_2 → 2AgCl + Mg(NO_3)_2$

GO! Activity 1.3.5

1. (a) 62·5 g (b) 241 g (c) 164 g (d) 400 g
2. (a) 28·1 g (b) 315·7 g (c) 139·4 g (d) 688·0 g
3. (a) 1·54 mol (b) 0·25 mol (c) 1·43 mol (d) 0·23 mol

GO! Activity 1.3.6

1. Distilled water has no dissolved chemicals in it.
2. Stir with a glass stirring rod.
3. All the solute has to be transferred to ensure the concentration of the solution is accurate.
4. To ensure all the solute is dissolved and mixed thoroughly.

255

GO! Activity 1.3.7

1. mol = mass/gfm = 0·25 mol, then c = mol/vol = 0·25/0·5 = 0·5 mol l^{-1}
2. mol = c x v = 0·2 × 0·15 = 0·03 mol
3. v = mol/c = 0·65/0·45 = 1·44 l
4. mol = c x v = 0·9 × 0·05 = 0·045 mol then, mass = mol × gfm = 3·6 g

GO! Activity 1.3.8

1. 72 g   2. 5·2 g

1.4 Acids and bases

GO! Activity 1.4.2

$SO_2(g) + H_2O(\ell) \rightarrow 2H^+(aq) + SO_3^{2-}(aq)$

GO! Activity 1.4.3

1. $CO_2 + 2NaOH \rightarrow Na_2CO_3 + H_2O$
2. Advantage = purer sodium hydroxide produced
   Disadvantage = risk of pollution from toxic mercury
3. Electricity is very expensive – expensive to dispose of the asbestos diaphragm safely.
   Sodium chloride is a relatively cheap resource and readily available. The hydrogen is not wasted – it is used as a source of energy in the process. The chlorine produced is an important chemical used to make other materials.
4. A = chlorine or hydrogen   B = hydrogen or chlorine
   X = electrolysis   Y = evaporation

GO! Activity 1.4.4

(a) pH5 (b) pH6

GO! Activity 1.4.5

1. $\dfrac{\text{(volume} \times \text{concentration) alkali}}{\text{balancing no. alkali}} = \dfrac{\text{(volume} \times \text{concentration) acid}}{\text{balancing no. acid}}$

$$\frac{23·6 \times 0·8}{1} = \frac{25 \times c}{1}$$

(Remember: you don't have to change volumes to litres)

$$18·88 = 25c$$

Concentration $= \dfrac{18.88}{25}$

$= \textbf{0·76 mol } l^{-1}$

2. 13.35 cm^3

Activity 1.4.6

(a) X = carbon dioxide

(b) $CuCO_3 + 2HCl \rightarrow CuCl_2 + H_2O + CO_2$

(c) No more bubbles of carbon dioxide would be seen.

Area 2: Nature's chemistry

2.5 Homologous series

Activity 2.5.1

Alkane	Number of carbons	Number of hydrogens	Molecular formula
nonane	9	20	C_9H_{20}
dodecane	12	26	$C_{12}H_{26}$
octadecane	18	38	$C_{18}H_{38}$
eicosane	20	42	$C_{20}H_{42}$

Activity 2.5.2

1. Although a line graph would clearly show the trend in boiling points, a spike graph should be used to plot the boiling points since points on the line between the main points plotted on a line graph have no meaning. You do not have alkanes with 1·5 carbon atoms, etc. Only the boiling point values at 1, 2, 3, etc. have meaning.

2. The trend is that the boiling point increases as the number of carbon atoms in the alkane molecules increases. The size of the increase from one member to another decreases as the molecules become larger.

   It would be predicted that the increase in boiling point from hexane to heptane would be slightly less than 33°C (the increase from pentane to hexane). The boiling point of heptane is 98°C (an increase of 31°C from hexane). The boiling point of octane is 126°C.

Activity 2.5.3

1. Structure 2: 2-methylpropane

2. 3-ethylpentane

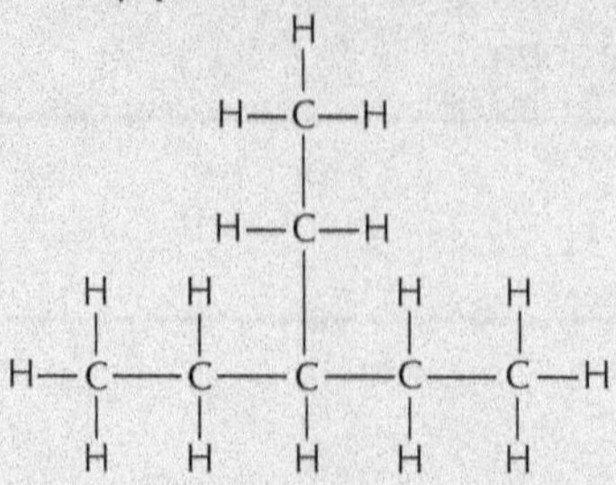

$CH_3CH_2CH(C_2H_5)CH_2CH_3$

2,4-dimethylpentane

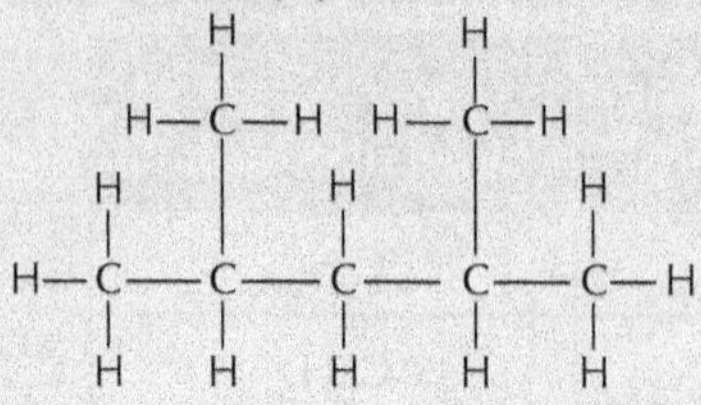

$CH_3CH(CH_3)CH_2CH(CH_3)CH_3$

3,3-dimethylpentane

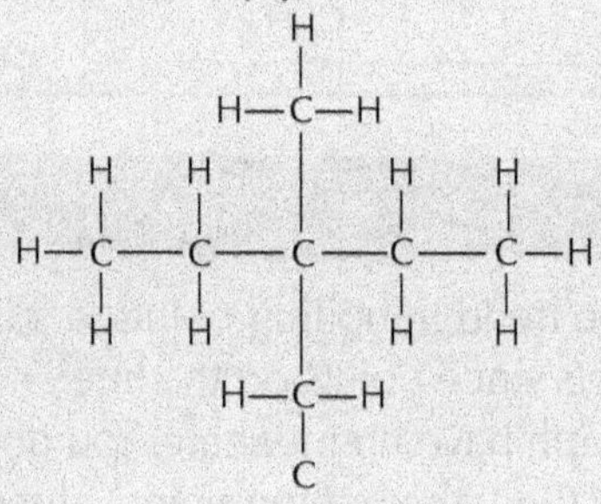

$CH_3CH_2C(CH_3)_2CH_2CH_3$

2,2,3-trimethylbutane

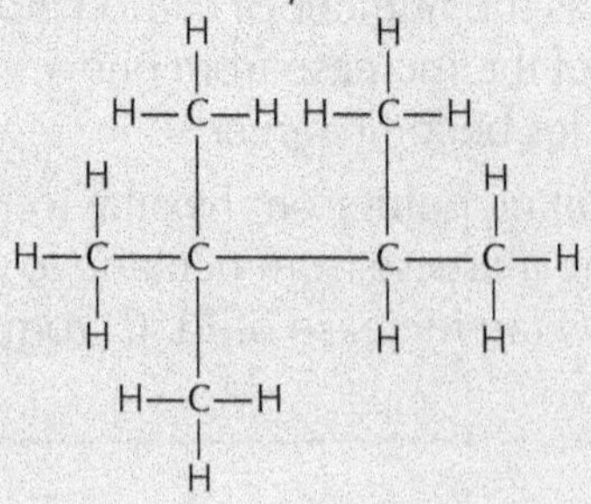

$CH_3C(CH_3)_2CH(CH_3)CH_3$

3.

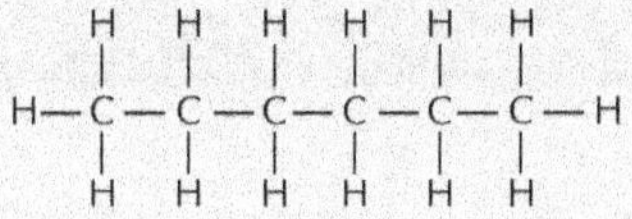

hexane

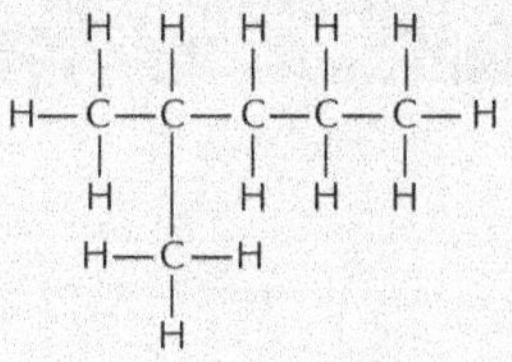

2-methylpentane

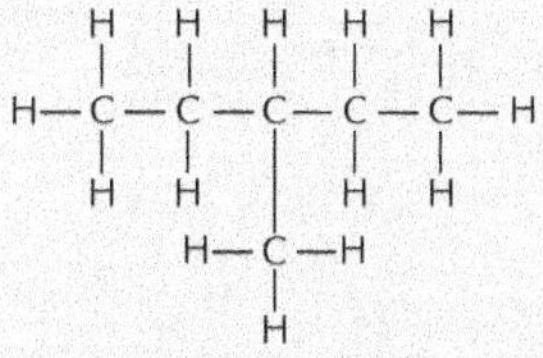

3-methylpentane

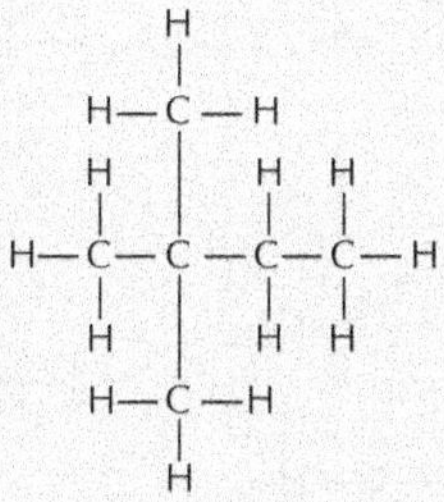

2,2-dimethylbutane

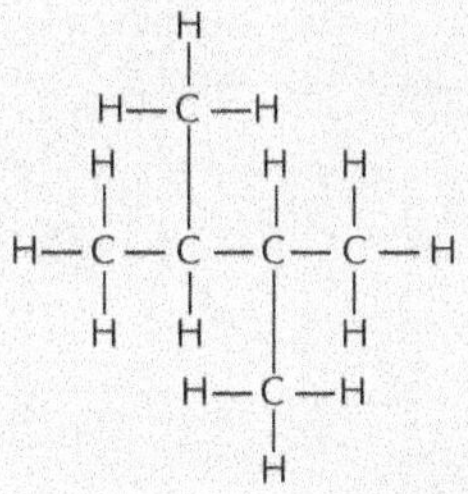

2,3-dimethylbutane

GO! Activity 2.5.4

Petrol used in winter needs to contain more volatile components than a summer blend. Short-chain hydrocarbons and branched-chain hydrocarbons increase volatility. In winter the petrol needs to be volatile enough to vaporise and mix with the air. Winter blends contain more low-boiling point alkanes such as butane and pentane. In summer less volatile components are used. If the blend is too volatile in summer, the petrol can vaporise too easily and cause a vapour lock that prevents petrol reaching the engine cylinders.

GO! Activity 2.5.5

1. (a) Cycloheptane

(b)

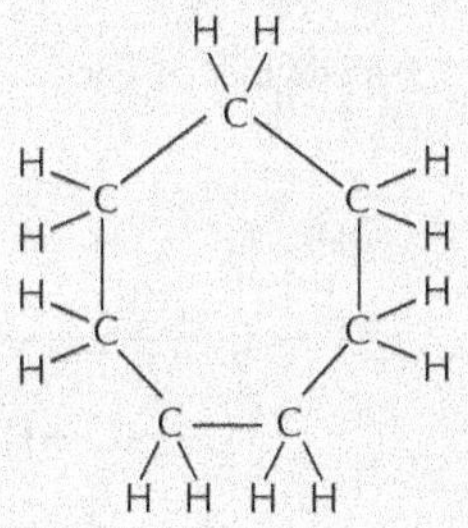

(c) Any alkene isomer of heptene.

2. (a) Cyclooctane

(b) C_8H_{16}

(c) Isomerism

Activity 2.5.6

1.

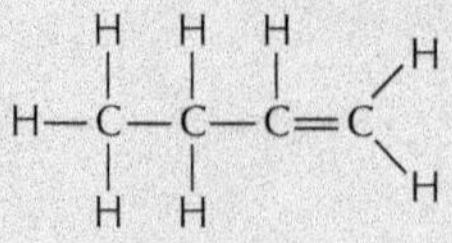

but-1-ene

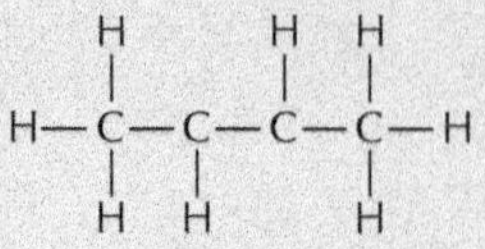

but-2-ene

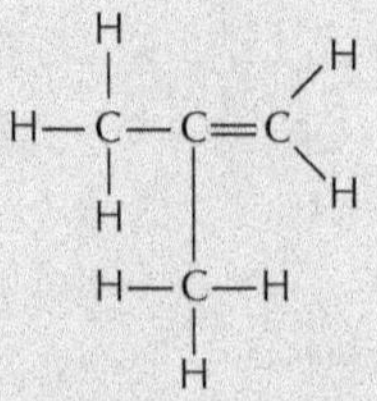

methylpropene

2. (a) 3-ethylpent-2-ene

(b) Shortened structural formula: $CH_3CH_2C(C_2H_5)CHCH_3$ or $CH_3CH_2C(CH_2CH_3)CHCH_3$ or $CH_3CHC(C_2H_5)CH_2CH_3$ or $CH_3CHC(CH_2CH_3)CH_2CH_3$

Molecular formula: C_7H_{14}

3.

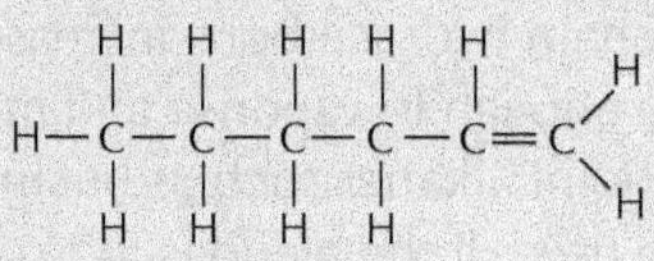

hex-1-ene

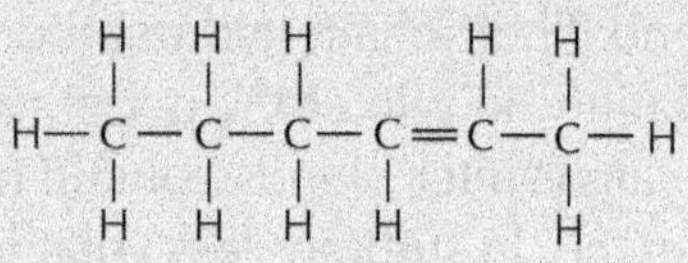

hex-2-ene

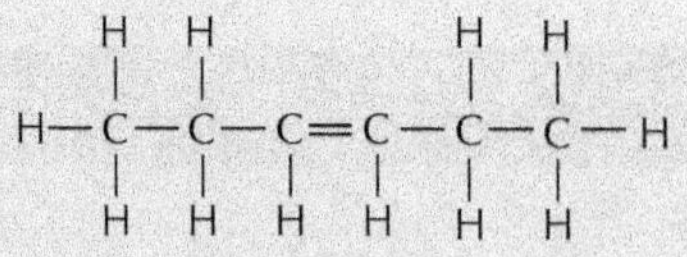

hex-3-ene

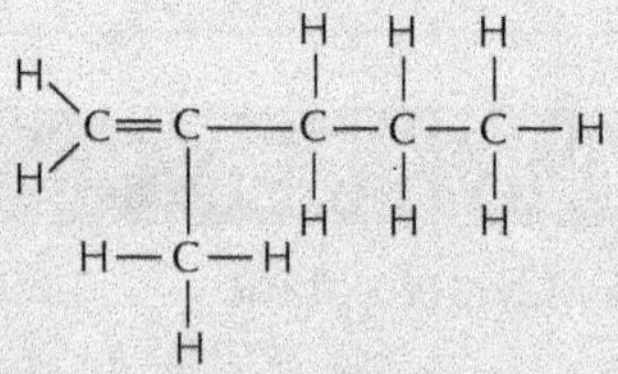

2-methylpent-1-ene

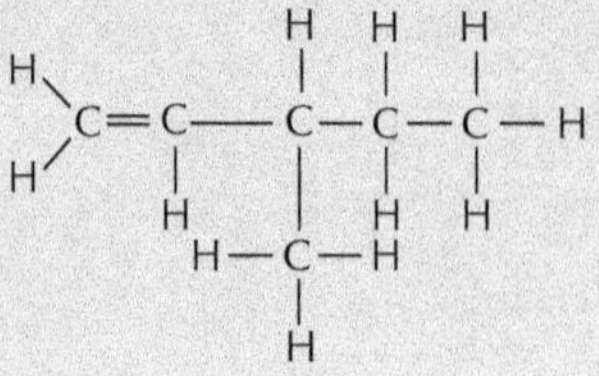

3-methylpent-1-ene

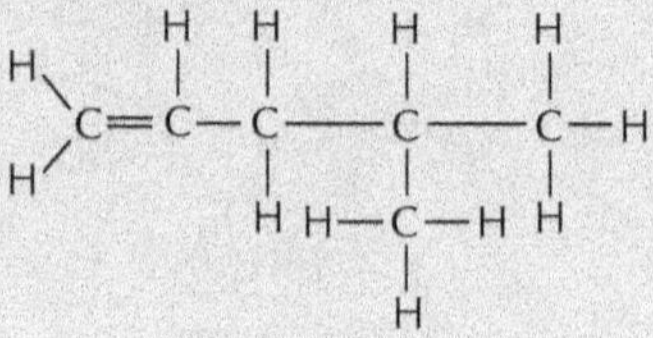

4-methylpent-1-ene

(continued)

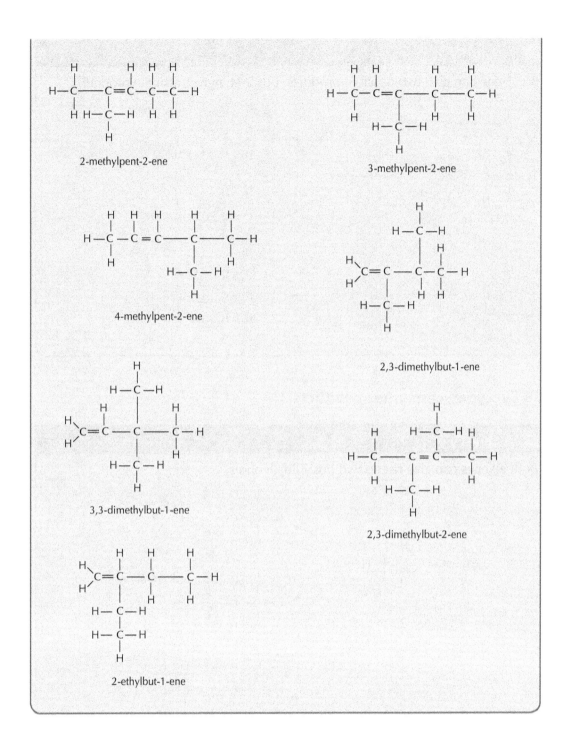

2-methylpent-2-ene

3-methylpent-2-ene

4-methylpent-2-ene

2,3-dimethylbut-1-ene

3,3-dimethylbut-1-ene

2,3-dimethylbut-2-ene

2-ethylbut-1-ene

GO! Activity 2.5.7

But-1-ene can give two possible products. But-2-ene gives a single product.

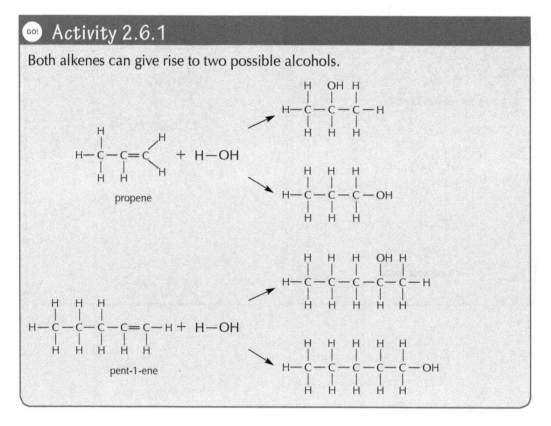

2.6 Everyday consumer products

GO! Activity 2.6.1

Both alkenes can give rise to two possible alcohols.

Activity 2.6.2

Alkane	Molecular formula	Alcohol	Molecular formula	Structural formula
methane	CH_4	methanol	CH_3OH	
ethane	C_2H_6	ethanol	C_2H_5OH	
propane	C_3H_8	propanol	C_3H_7OH	
butane	C_4H_{10}	butanol	C_4H_9OH	
pentane	C_5H_{12}	pentanol	$C_5H_{11}OH$	
hexane	C_6H_{14}	hexanol	$C_6H_{13}OH$	
heptane	C_7H_{16}	heptanol	$C_7H_{15}OH$	
octane	C_8H_{18}	octanol	$C_8H_{17}OH$	

GO! Activity 2.6.3

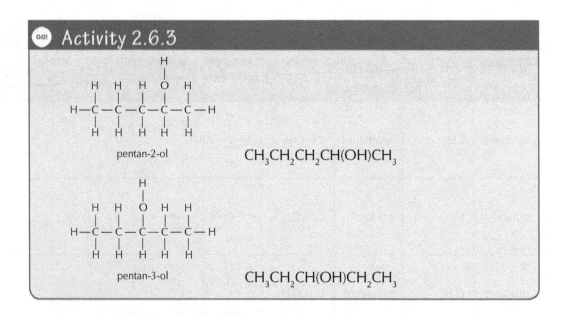

pentan-2-ol

$CH_3CH_2CH_2CH(OH)CH_3$

pentan-3-ol

$CH_3CH_2CH(OH)CH_2CH_3$

GO! Activity 2.6.4

1. (a)

Alcohol	Shortened structural formula	Melting point/°C
hexan-1-ol	$CH_3(CH_2)_4CH_2OH$	−52
heptan-1-ol	$CH_3(CH_2)_5CH_2OH$	−34
octan-1-ol	$CH_3(CH_2)_6CH_2OH$	−16

(b) (i)

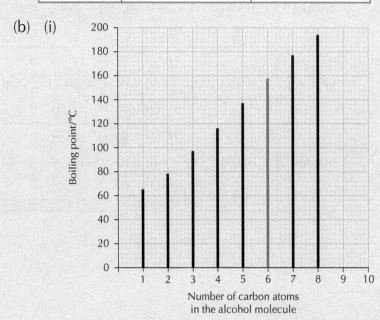

(ii) Predicted boiling point of hexan-1-ol: 157°C

Activity 2.6.5

(a) methanol

$$2CH_3OH(\ell) + 3O_2(g) \rightarrow 2CO_2(g) + 4H_2O(\ell)$$
(or multiples of the equation)

(b) propan-1-ol

$$2C_3H_7OH(\ell) + 9O_2(g) \rightarrow 6CO_2(g) + 8H_2O(\ell)$$

Activity 2.6.6

Advantages include:

- Ethanol is a renewable fuel that can be obtained from plant materials whereas petrol is obtained from crude oil, a fossil fuel and finite resource.
- Can reduce a country's dependence on imported oil.
- Ethanol burns very cleanly giving carbon dioxide and water. Petrol emissions include unburned hydrocarbons and soot.
- Greenhouse gas emissions are lower.
- Cost of converting vehicles to run on ethanol is low.

Disadvantages include:

- Land that could be used to grow food crops is being used to grow plants which are turned into fuel.
- Large amounts of fertilisers are required to sustain crop yields.
- Diesel from fossil fuels is used to run machinery needed to harvest and process crops to produce ethanol.
- Ethanol is expensive to produce.
- Burning ethanol gives less energy than petrol meaning more ethanol needs to be burned to drive the same distance.
- There is limited availability of ethanol.
- Ethanol can absorb water on standing, which can affect engine wear.

Activity 2.6.7

1. (a) 9.82 kJ   (b) 10.1 kJ
2. 2.4
3. 200 g
4. 12.2°C

Activity 2.6.8

1. (a) butanoic acid

   (b) heptanoic acid

2. (a) pentanoic acid

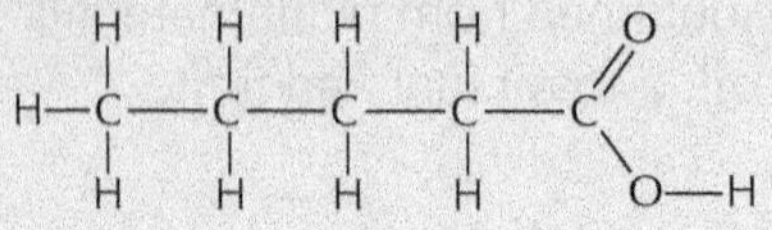

   octanoic acid

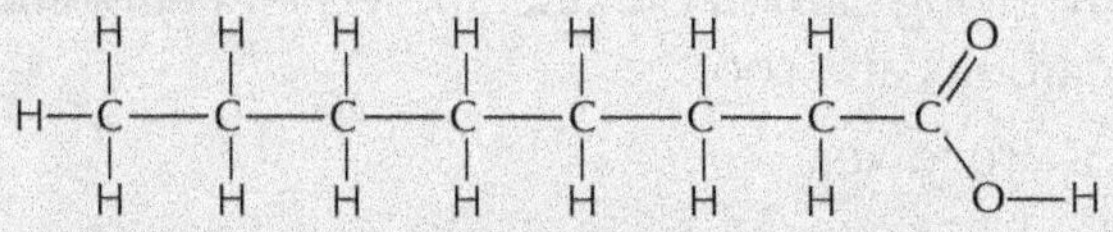

   (b) A C_4H_9COOH B $C_7H_{15}COOH$

3. (a) ethanoic acid + zinc → zinc ethanoate + hydrogen

   $2CH_3COOH + Zn \rightarrow (CH_3COO)_2Zn + H_2$

   (b) butanoic acid + potassium oxide → potassium butanoate + water

   $C_3H_7COOH + K_2O \rightarrow C_3H_7COOK + H_2O$

   (c) propanoic acid + lithium hydroxide → lithium propanoate + water

   $C_2H_5COOH + LiOH \rightarrow C_2H_5COOLi + H_2O$

   (d) methanoic acid + copper(II) carbonate → copper methanoate + water + carbon dioxide

   $HCOOH + CuCO_3 \rightarrow (HCOO)_2Cu + H_2O + CO_2$

4. Preservatives, manufacture of soap, and some medicines.

Area 3: Chemistry in society

3.7 Metals

GO! Activity 3.7.1

Percentage of copper in malachite = 57·5%

GO! Activity 3.7.2

1. $2CuO(s) + C(s) \rightarrow 2Cu(s) + CO_2(g)$
2. $Cu^{2+}(s) + 2e^- \rightarrow Cu(s)$

GO! Activity 3.7.3

1.

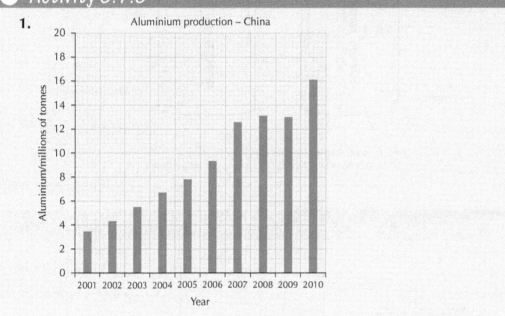

Aluminium production – China

2. Greater than 20 million tonnes (actually estimated at 31.5 million tonnes)
3. 2·5 million tonnes (to 1 decimal place) of alumina used each month

GO! Activity 3.7.4

1. (a) $Mg(s) \rightarrow Mg^{2+}(s) + 2e^-$ oxidation
   (b) $O_2(g) + 4e^- \rightarrow 2O^{2-}(s)$ reduction
2. $2Mg(s) + O_2(g) \rightarrow 2Mg^{2+}(s) + 2O^{2-}(s)$ redox

GO! Activity 3.7.5

1. (a) $Na(s) \rightarrow Na^+(aq) + e^-$ oxidation
   (b) $2H_2O(\ell) + 2e^- \rightarrow H_2(g) + 2OH^-(aq)$ reduction
2. $2Na(s) + 2H_2O(\ell) \rightarrow H_2(g) + 2Na^+(aq) + 2OH^-(aq)$ redox

GO! Activity 3.7.6

1. (a) $Fe(s) \rightarrow Fe^{2+}(aq) + 2e^-$ oxidation
   (b) $2H^+(aq) + 2e^- \rightarrow H_2(g)$ reduction

2. $Fe(s) + 2H^+(aq) \rightarrow Fe^{2+}(aq) + H_2(g)$ redox

GO! Activity 3.7.7

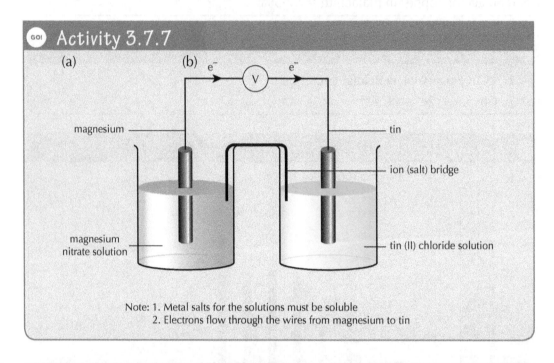

Note: 1. Metal salts for the solutions must be soluble
   2. Electrons flow through the wires from magnesium to tin

GO! Activity 3.7.8

1.

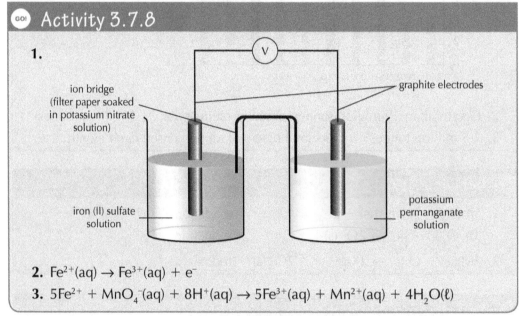

2. $Fe^{2+}(aq) \rightarrow Fe^{3+}(aq) + e^-$

3. $5Fe^{2+} + MnO_4^-(aq) + 8H^+(aq) \rightarrow 5Fe^{3+}(aq) + Mn^{2+}(aq) + 4H_2O(\ell)$

3.8 Plastics

GO! Activity 3.8.1

Plastic	Use
poly(ethene)	hockey pitch surface
polyurethane	adhesive
synthetic rubber	hockey pitch base
PVC	tents at shooting venue, panels in structure of main stadium, aquatic centre wrap, surface of velodrome
polyester	panels in main stadium, athletes' uniforms, football boots
carbon fibre-reinforced plastic	bicycle frames
PET (recyclable polyester)	beer bottles

GO! Activity 3.8.3

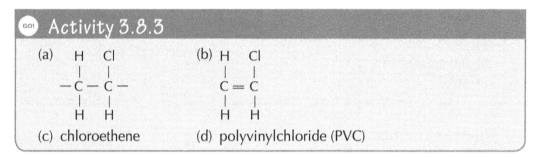

(a)   (b)

(c) chloroethene   (d) polyvinylchloride (PVC)

3.9 Fertilisers

GO! Activity 3.9.1

(a) and (b)

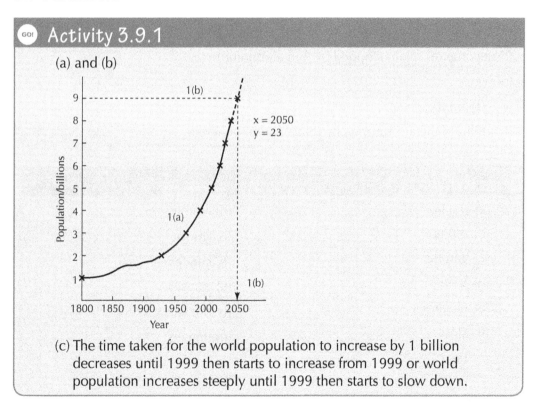

$x = 2050$
$y = 23$

(c) The time taken for the world population to increase by 1 billion decreases until 1999 then starts to increase from 1999 or world population increases steeply until 1999 then starts to slow down.

Activity 3.9.2

(a) World population increasing, so more food needs to be produced, so more fertiliser needed.

(b) N: 112; P: 43; K: 37 (all values approximate).

(c) Increases, because world population continues to increase.

Activity 3.9.3

1. (a) Largest producer
   (b) Not one of the top ten exporters.
   (c) China uses most of the ammonia it produces.

2. (a) Biggest exporter
   (b) Natural gas
   (c) USA, because it has a large population and so needs to grow a lot of crops. Geographically close.

3. (a) 200°C
   (b) It increases
   (c) 60%

3.10 Nuclear chemistry

Activity 3.10.1

(a) α (alpha)

(b) 2+

(c) few metres in air; stopped by thin aluminium

(d) 1−

(e) γ (gamma)

(f) miles in the air; stopped by thick lead or concrete

Activity 3.10.3

2. (a) nuclei
   (b) alpha
   (c) gamma
   (d) helium
   (e) electrons
   (f) slow
   (g) beta
   (h) aluminium
   (i) smoke
   (j) cancer
   (k) ionises
   (l) thickness
   (m) tracers

Activity 3.10.4

1. a = 216 b = 85 X = At

2. $^{234}_{92}U \rightarrow ^{4}_{2}He + ^{230}_{90}Th$

3. $^{228}_{89}Ac \rightarrow ^{0}_{-1}e + ^{228}_{90}Th$

Activity 3.10.5

1.

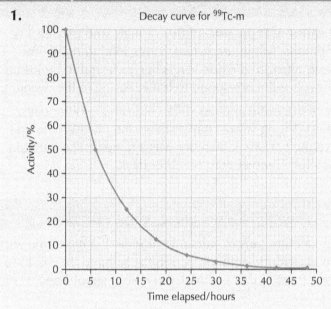

Decay curve for ^{99}Tc-m

2. A long half-life means that it will not stop emitting α particles in the lifetime of the smoke detector, which is essential for safety reasons.

Activity 3.10.6

When an electron collides with the potassium nucleus a proton is changed to a neutron. The atomic number of the radioisotope decreases by 1 unit. The nucleus changes from a potassium isotope to an argon isotope. The transition is accompanied by the emission of a gamma radiation.

Activity 3.10.7

$^{87}_{37}Rb \rightarrow ^{87}_{38}Sr + ^{0}_{-1}e$

3.11 Chemical analysis

GO! Activity 3.11.1

(a) 56·60%
(b) 49·8 tonnes per year
(c) 18·35%
(d) 141·4 tonnes per year

GO! Activity 3.11.2

The nitrogen dioxide level varies with the time of day or night and may also be influenced by weather conditions. Minimum values correspond to night time when traffic is low. The rise in nitrogen dioxide levels begins as traffic builds up in the morning. The fall in levels will correspond to traffic easing in the evening.

The pattern for the first day of the cycle appears different from the other days. This could be weather-related. Strong winds may disperse the nitrogen dioxide or heavy rain might dissolve the gas. 30 September was a Sunday. The nitrogen dioxide peak on that day is slightly lower than on other days.

GO! Activity 3.11.3

A flame test would confirm calcium ions in calcium chloride. If a nichrome wire is cleaned using 4 mol l^{-1} hydrochloric acid, dipped in the salt and then held to the side of the blue cone of a Bunsen flame a brick red colour should be observed.

Adding silver nitrate solution to a solution of calcium chloride will give a white precipitate indicating the presence of chloride ions.

Answers to exam-type questions

Area 1: Chemical changes and structure

1.1 Rates of reaction

1. (a) So that none of the reaction mixture spray can escape.

(b) (i) 0·017 (g s^{-1})

(ii) 0·39 (g)

2. (a) 0·0015 mol l^{-1} min^{-1}

(b) New line should start at same point as original and should have a steeper gradient.

3. (a) (i) 2·75 litres per microsecond

(ii) 4·5 microseconds

4. (a)

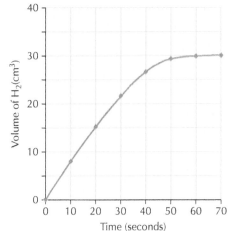

(b)

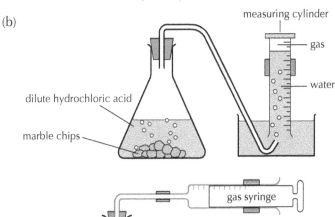

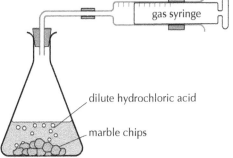

1.2 Atomic structure and bonding

1. D
2. A
3. B
4. (a) 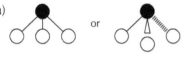 (b)

5. (a) covalent network
   (b) (i) 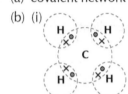 (ii)

$$\begin{array}{c} H \\ | \\ H - C - H \\ | \\ H \end{array}$$

6. (a) Network
   (b) Sb_2O_3
   (c) (i) $^{11}_{5}B$
   (ii) isotopes
7. (a) Covalent network (b) ionic lattice (c) metallic lattice (d) covalent molecular
   (b) SiO_2

1.3 Formulae and reaction quantities

1. A
2. B
3. B
4. (a) $2NaN_3(s) \rightarrow 2Na(s) + 3N_2(g)$
   (b) $2H_2O_2(aq) \rightarrow O_2(g) + 2H_2O(\ell)$
5. (a) 0·02 mol
   (b) 0·1 mol
6. 0·5 mol
7. 1·62 g

1.4 Acids and bases

1. B
2. A
3. B
4. A
5. (a) C
   (b) Any number above 4·4 but below 6
6. (a) (i) Red, pink, orange, yellow
      (ii) Line must be increasing
          Line stops at pH7 or below
   (b) 0·005 mol

7. Concentration of Na_2CO_3 solution = 0·112 mol l^{-1}

8. (a) An indicator such as methyl orange must be added.

   (b) Concentration of KOH(aq) = 0·123 mol l^{-1}

Area 2: Nature's chemistry

2.5 Homologous series

1. C (An alkane, both obey general formula C_nH_{2n+2})

2. C (Both have molecular formula C_5H_{12})

3. A

4. (a) Dehydrogenation or cracking

   (b)

   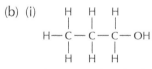

   (c) Pentane and 2-methylbutane have same molecular formula. The sequence of reactions has led to a change in the structure. The process has therefore produced an isomer of pentane, hence isomerisation.

   (d) Branched-chain alkanes are blended into petrol to improve the octane rating and reduce engine knock.

5.

6. (a) 2,3-dimethylbutane

   (b)

7. (a) As the number of carbons in the straight-chain alkane increases the critical temperature increases.

   (b) 220–225°C (approximately 10–15° below critical temperature of straight-chain isomer)

2.6 Everyday consumer products

1. (a) hydroxyl group

   (b) (i)

   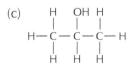

   (ii) hexan-1-ol is the largest molecule and will have the strongest intermolecular forces. It therefore has the highest boiling point.

   (iii) The boiling point of octan-1-ol is 195°C. Based on the figures for the other alcohols it is reasonable to predict a boiling point of 197–199°C.

   (c)

2. (a) (i) Homologous series

(ii) $C_nH_{2n+2}S$

(b) Sulfur dioxide (SO_2)

(c)

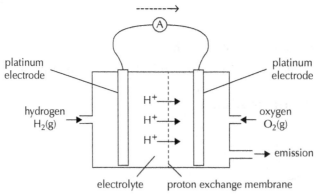

or $CH_3CH(SH)CH_3$

3. (a)

or $CH_3CH_2CH_2CH(OH)CH_3$

(b) (i) If the functional (hydroxyl) group is on an end carbon the energy released from the alcohol will be greater.

(ii) 3971 kJ (an answer in the range 3968–3971 kJ would be acceptable)

4. (a)

(b) (i) As the number of carbon atoms in the carboxylic acid increases the boiling point increases.

(ii) 181–190°C

Area 3: Chemistry in society

3.7 Metals

1. (a) (i) $2H_2 + O_2 \rightarrow 2H_2O$

(ii) From left to right through the wire

(b) Only emission from fuel cell is water whereas burning petrol releases carbon dioxide and other gases into the atmosphere.

Energy is being obtained from a renewable source whereas petrol is obtained from crude oil, a finite resource.

2. (a) $2CH_3OH(\ell) + 3O_2(g) \rightarrow 2CO_2(g) + 4H_2O(\ell)$

(b) The methanol fuel cell reaction produces carbon dioxide.

3.

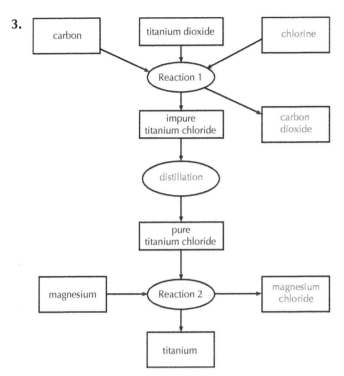

4. (a) Mass of titanium = 3·6 g

   (b) Moles of titanium = 0·075 mol

3.8 Plastics

1. B

2. D

3. (a) hydroxyl

   (b)

```
      H   H
      |   |
   — C — C —
      |   |
      H   OH
```

   (c) soluble/dissolves in water

4. (a) Carbon to carbon double bond

   (b)

```
      H      H   H      H   H      H
      |      |   |      |   |      |
   ~ C —— C — C —— C — C —— C ~
      |      |   |      |   |      |
    C6H5   H   C6H5  H   C6H5   H
```

   (c) poly(phenylethene)

5.

```
   H   COOCH3  H   COOCH3  H   COOCH3
   |   |       |   |       |   |
 — C — C ————— C — C ————— C — C —
   |   |       |   |       |   |
   H   CN      H   CN      H   CN
```

6. (a) man-made/made in a factory

   (b) (i)

```
     H   CH3      H   CH3      H   CH3
     |   |        |   |        |   |
   — C — C —————— C — C —————— C — C —
     |   |        |   |        |   |
     H   COOCH3   H   COOCH3   H   COOCH3
```

   (ii) addition

3.9 Fertilisers

1. (a) $2KOH(aq) + H_2SO_4(aq) \rightarrow K_2SO_4(aq) + 2H_2O(\ell)$

   (b) neutralisation

   (c) $(78/174) \times 100 = 44 \cdot 8\%$

   (d) $(NH_4^+)_3PO_4^{3-}$

2. (a) (i) Ostwald

   (ii) reaction is exothermic/gives out heat

   (iii) water

   (b) % = $(28/80) \times 100 = 35$

3. (a) lightning

   (b) ammonia

   (c) As the temperature increases solubility decreases.

4. (a)

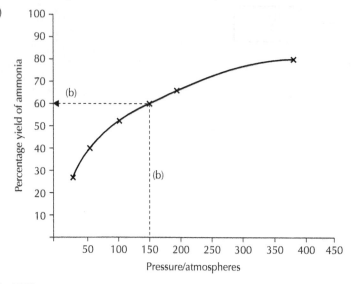

   (b) 60%

   (c) base or alkali

5. (a) $Ba(OH)_2 + 2NH_4Cl \rightarrow 2NH_3 + BaCl_2 + 2H_2O$

   (b) moist pH paper turns blue

   (c) water freezes

3.10 Nuclear chemistry

1. D

2. C

3. C

4. A

5. (a) $^3_1H \rightarrow {}^3_2He + {}^0_{-1}e$

   (b) (i) rate of formation = rate of decay

   (c) 1/8 = 3 half-lives; $3 \times 12 \cdot 3 = 36 \cdot 9$ years

6. (a) $_{-1}^{0}$e (beta particle)

(b) (i) draw a graph with a curve that passes down through points:

time/hours: 0   6   12   18   24

    mass/g: 0·5   0·25   0·125   0·06   0·03

(ii) short half-life (doesn't remain active in body for a long time)

7. (a) $_{38}^{89}$Sr → $_{39}^{89}$Y + $_{-1}^{0}$e

(b) (i) no effect
(ii) $(89/160) \times 10 = 5·56$ g

(c) ¼ (0·25)

8. (a) Gamma rays are produced when the nucleus emits excess energy.

(b) Coal between the gamma source and the detector will absorb some of the gamma radiation lowering the reading on the detector. When coal falls below the level of a detector the reading on the detector will increase indicating an increase in radiation reaching the detector.

(c) Gamma radiation is the most penetrating type of radiation. Since the source and detector would need to be on the outside of the hopper, radiation would need to pass through the sides of the hopper. α and β radiations would be unable to pass through the sides of the hopper.

9. (a)

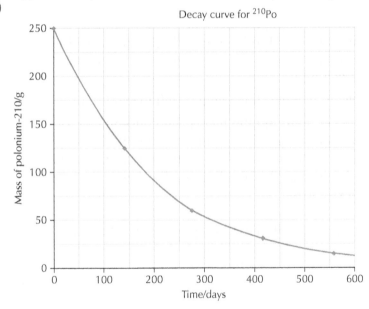

Decay curve for ^{210}Po

(b) $_{84}^{210}$Po → $_{82}^{206}$Pb + $_{2}^{4}$He

3.11 Chemical analysis

1. D

2. B

3. (a) Step 2: Add a few drops of indicator.
    Step 6: Note the volume of acid at the end-point.
(b) Pipette.
(c) Titration.
(d) $\dfrac{(19.3 + 19.5)}{2} = 19.4$ cm^3

Open-ended and extended answer questions and assignment: answers

Sample answer to open-ended Example 1

Iodine-131 is used to detect and treat cancer. It emits radiation, which is α or β particles or γ rays. α particles are helium nuclei and β particles are high energy electrons. γ rays are very high energy and can travel over a long distance. Radiation ionises molecules they come in contact with. There is the possibility of damage to human tissue if these emissions get into the body. The health workers may be worried that the iodine could be on the linen and it could get into their body in some way.

(You could also have included the possibility of the iodine having a long half-life and so could be active for a long time. Nuclear equations could have been included to show ways that iodine might decay.)

Sample answer to open-ended Example 2

Acids contain hydrogen ions and hydroxide ions. There is a higher concentration of hydrogen ions. Alkalis contain hydroxide ions and hydrogen ions. There is a higher concentration of hydroxide ions. Pure water contains equal concentrations of hydrogen and hydroxide ions.

This is another possible answer which would gain full marks.

All acids contain hydrogen ions and the higher the concentration of the hydrogen ions the lower the pH. There are also hydroxide ions present but the concentration is lower than the hydrogen ion concentration. Alkalis also contain hydrogen ions but the concentration is lower than hydroxide ions. The lower the concentration of hydrogen ions, the higher the pH.

Sample answer to extended response Example

α particles don't travel far through the air and have low penetration power (stopped by a sheet of paper) so will not get through the casing of the detector.